The Essential Bonsai Book

JONAS DUPUICH
Photography by David Fenton

The Essential Bonsai Book

Techniques for Creating Beautiful Trees

TEN SPEED PRESS
California | New York

Trident maple

CONTENTS

Chinese wisteria 'Caroline'

INTRODUCTION

A few years ago, a student who was moving to another country asked if I could create a guide to help him grow bonsai in a different climate. It was an interesting request. I knew I wouldn't always be available to help when questions came up, so I wanted to provide a way for him to work independently and learn from any failures or successes. The resulting conversations became the core of this book.

Who This Book Is For

The Essential Bonsai Book is for enthusiasts who are taking steps to improve their collections. Whether you've found a teacher to work with or are learning on your own, *The Essential Bonsai Book* provides the essentials you need to get started so you can avoid common traps that slow development.

Unlike other bonsai books, *The Essential Bonsai Book* doesn't just tell you what to do. Instead, it will help you set goals for your collection and provide the information you need to reach them.

When someone tells you what to do with your trees, think twice: what works for them may not work for you. You know more about your garden than anyone else. You know what soil your trees grow in and how often they receive water and fertilizer. You know the climate your trees grow in and what your goals are for their development. *The Essential Bonsai Book* will help you leverage this experience so you can build a better collection, feel better about the work you do on your trees, and get more enjoyment out of this wonderful hobby.

About Me

I started growing bonsai more than thirty years ago when I met Boon Manakitivipart (see profile on page 195). Within a few months, I had a small collection of trees and a batch of young pines. Boon had encouraged me to start the pines from seed as he knew I liked the species but had trouble finding good material to work with. This proved to be a great way to learn the ins and outs of the species and understand how to work with bonsai at different stages of development.

In time, I started sharing my experiences on the *Bonsai Tonight* blog, and in 2020, I wrote *The Little Book of Bonsai*, a guide that introduces readers to bonsai and helps them get started with the hobby.

From the beginning, my goal for teaching and writing about bonsai has been to help as many people as possible appreciate the beauty of bonsai. Toward this end, I began working with bonsai full time in 2015, and in 2022 I cofounded, with Eric Schrader, the Pacific Bonsai Expo, a biennial exhibit featuring outstanding trees from across the United States.

How This Book Is Organized

The Essential Bonsai Book provides an overview of the work that goes into creating a beautiful bonsai collection.

The first three chapters offer a framework for building your collection. **Chapter 1** provides tips for expanding (or refining) your collection and for selecting trees that are well-suited to the bonsai work you like best. **Chapter 2** focuses on bonsai design to help you create trees that look the way you want them to look. **Chapter 3** helps you set goals for your trees and identify techniques that are appropriate for different stages of development.

The next three chapters provide how-to guides for working on your trees. **Chapter 4** describes the fundamental techniques of pruning and wiring. **Chapter 5** describes more advanced training techniques such as pinching and defoliating that can help you create beautiful branch structures. **Chapter 6** offers an in-depth overview of how, when, and why we repot bonsai.

Chapters 7 and 8 focus on bonsai care. **Chapter 7** helps you identify trees that grow well in your climate and offers strategies for keeping bonsai healthy through summer heat and winter cold. **Chapter 8** focuses on watering and fertilizing to help you water effectively and provide your trees with the nutrients they need to produce new growth and maintain health.

The last two chapters feature case studies. **Chapter 9** provides case studies to help you identify when to apply given techniques to your trees and shows you what the work looks like. **Chapter 10** is an in-depth case study that breaks down the complete styling of a shimpaku juniper over the course of one year.

At the end of selected chapters, you'll find tips from good friends who are respected bonsai professionals. I've learned a lot from this group over the years, and these vignettes showcase the insights I've found most helpful.

Finally, in terms of the kinds of trees covered in *The Essential Bonsai Book*, I've included a mix of popular bonsai species such as junipers, pines, and maples, as well as less-commonly grown species like cryptomeria and coast redwood. I chose this mix because they're species with which I'm familiar and because they allow me to demonstrate a greater diversity of techniques that may prove helpful as you experiment with species that grow well where you live.

Terms Used in This Book

Tree-Related Terms

apex—the top, or crown, of the tree

collected tree—a tree that grew in the natural landscape and was dug up to be trained as bonsai (*yamadori* in Japanese)

flow—the general movement of a tree from left to right or right to left

front—the side of the tree oriented toward the viewer; bonsai are styled to look their best when viewed from the front

literati bonsai—trees with few branches, subtle trunk movement, and slender trunks (*bunjin* in Japanese)

movement—the curves in a branch or trunk; a tree is said to have "good movement" when the trunk has interesting curves or twists that are appropriate for the style of the tree

pre-bonsai—a tree grown for bonsai that does not yet fully look like a bonsai (for example, a tree with a good trunk but little to no branch development)

refinement—a stage of development in which a tree has the basic form of a bonsai and the training goal is to make relatively subtle improvements over time (to improve, for example, the character, density, and arrangement of the branches)

shohin—small bonsai less than 8 inches tall

silhouette—the outline of the canopy; used to describe the shape of the tree

surface roots—major roots growing near the surface of the soil at the base of the trunk (*nebari* in Japanese)

Branch-Related Terms

back bud—a bud that forms along the length of a branch (not at the tip)

bifurcation—when a branch (or trunk or root) divides into two smaller branches (or trunks); a bifurcating branch pattern is the fundamental pattern for creating branch density in bonsai

branch pad—a block of foliage typically formed by a primary branch and its branchlets

candle—new shoots or buds on pines that elongate before producing needles

harden off—when the leaf cuticle has completely formed (when a leaf becomes mature)

interior branches—branches that are sheltered by the outer canopy

internode—the space between nodes along a trunk or branch

key branch—the largest or most prominent branch on the tree; a branch that helps determine the tree's flow

leader—the main branch in a branch pad; the main or longest trunk

node—a point on a branch or trunk from which buds may develop

primary branches—branches that grow from the trunk

sacrifice branch—a branch used to thicken part of the tree (typically the trunk or a prominent branch) that you remove when the trunk (or branch) reaches the desired thickness

secondary branches—side branches that grow from primary branches

terminal bud/apical bud—a bud at the end of a shoot or branch

tertiary branches—side branches that grow from secondary branches

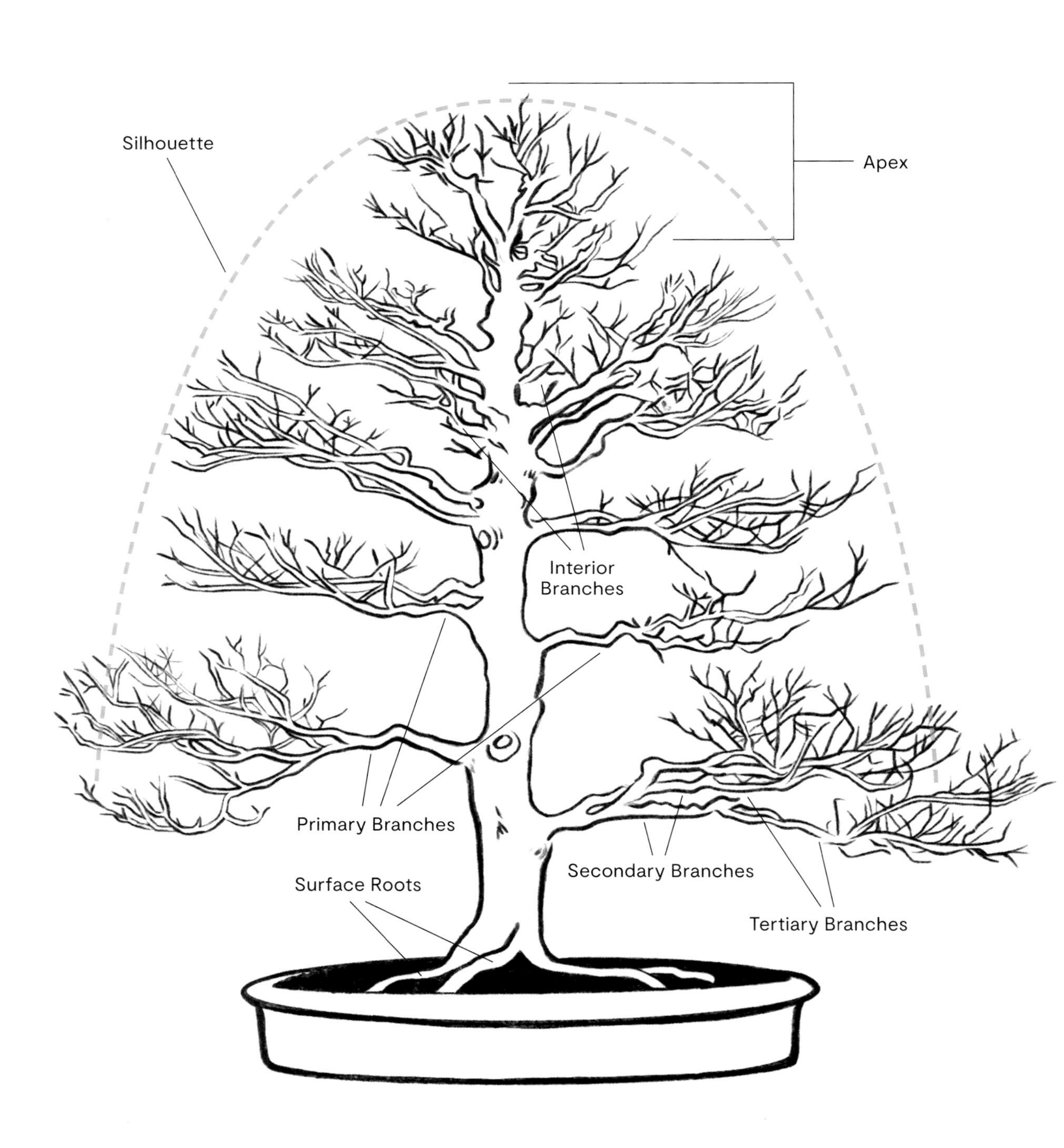
Silhouette
Apex
Interior Branches
Primary Branches
Secondary Branches
Surface Roots
Tertiary Branches

Terms Related to Techniques

air layering—a propagation technique that entails stimulating roots on a trunk or branch and then separating the rooted section of the tree to create a new tree

bare rooting—an approach to repotting in which all of the soil is removed from the root ball so it can be replaced with new soil

cutback—a form of pruning in which the goal is to define or maintain the shape of the tree or to stimulate budding

cut paste—compounds applied to wounds (caused by pruning or otherwise) to stimulate callus production, slow the decay of exposed wood, and make wounds less visible over time

decandling—a technique for removing spring growth (candles) on selected pine species to stimulate compact growth and shoot bifurcation

defoliation—a technique for removing some (partial defoliation) or all (full defoliation) of the leaves on a tree to slow growth and allow light into the tree's interior

leaf cutting—reducing the size of a leaf by removing a portion of it with scissors

scion—a detached branch or shoot used for grafting

side-veneer grafting—a form of grafting in which a scion is inserted where a new branch is desired

thinning—a form of pruning for decreasing foliar density without changing the shape of the tree

thread grafting—a form of grafting in which the end of a branch (that is still attached to the tree) is "threaded" through a hole drilled in the trunk where a new branch is desired

Terms Related to Deadwood Features

deadwood/deadwood features—areas of the tree, generally sections of the trunk or branches, that are no longer alive; deadwood features are often preserved on bonsai to convey the tree's response to harsh growing conditions

jin—dead branches from which the bark has typically been removed

lifeline—the living areas of the trunk (generally the cambium, phloem, and bark layers)

shari—dead areas of the trunk from which the bark has typically been removed

CHAPTER 1

Building a Quality Collection

What is a "quality collection" and why would you want to create one? Isn't it enough to maintain your trees as they are?

A quality collection is a group of healthy and—in your eyes—beautiful trees. It's a collection that fits your lifestyle (in terms of budget, space, and free time) and supports your goals for growing bonsai.

Any number of experiences can inspire you to create and maintain a quality collection. Maybe you want to develop bonsai that have the same characteristics as trees you admire in nature. Maybe you want to enter a tree in a local show. Or maybe you want to create the most beautiful trees possible.

The first time I saw truly outstanding bonsai was during a visit to Japan. Before then, I'd seen many trees that in time would become great, and I'd seen great work on trees of differing quality. But it was in Japan that I first saw the effect of great work done to great trees over long periods of time. To this day, I remember the twiggy branches and flowers that glowed orange on dwarf flowering quince and the deep green needles on black pines that seemed impossibly dense and even. Ever since that trip, I've worked to create trees with similar features in my own garden.

CHARACTERISTICS OF QUALITY BONSAI

One of the first steps toward building a quality collection is recognizing what excites you most about bonsai. Some people prefer large trees with powerful deadwood features and craggy bark. Others prefer delicate trees with graceful movement. In general, experienced bonsai growers appreciate trees that convey great age, superior levels of refinement, and unique character.

A great deal of this character resides in the trunk. When you learn to evaluate the trunk apart from how it's oriented in the pot or whether or not it has good branches, you can appreciate the bulk of the tree's value and better understand its potential.

A natural next step is to visualize what you can do with the rest of the tree. The more clearly you can visualize this, the more successfully you can create the designs you have in mind for your trees.

A good way to learn how to evaluate specific parts of a tree is by asking a series of questions about it. Once you have the answers, you'll find it much easier to determine whether or not a tree can help you reach your goals for your collection.

Evaluating the Trunk

The following questions can help you identify good and bad points in the trunk.

- Is the movement interesting? Is there something unique or unusual about the line of the trunk?
- Is the movement appropriate for the species or style of the tree you're looking to make?
- Is the taper attractive, and do the transitions of taper look graceful or artificial?
- Are there unsightly scars on the trunk? And if so, are they on the front or the back of the tree?
- Are there deadwood features on the trunk, and are they appropriate for the species?
- Are there opportunities for creating deadwood features that would improve the look of the trunk?

You can also consider characteristics that are highly prized in the species you are growing. For junipers, look for twisting movement in the trunk and compelling deadwood features. For pines, look for deep plates of aged bark. For deciduous species, look for aged trunks that are free of scars. And for broadleaf evergreen species such as oak or olive, look for a trunk with good bark and/or interesting deadwood features.

Evaluating Surface Roots

The area of a bonsai where the trunk transitions to roots demonstrates the tree's connection to the soil in which it grows. The nature of this connection, whether stable or tenuous, can have a great effect on the look of a tree.

- Does the lower trunk transition gracefully to the surface roots?
- Do roots emerge radially from the base of the trunk or are there awkward gaps between them?
- Are the roots roughly the same size?
- Are the roots too small (which can make a tree look young) or too large (which can make it look clunky)?
- Will the roots fit into a bonsai pot?
- Are there opportunities for improving the roots by pruning, grafting, or layering?

Roots matter more for deciduous trees than for most conifers, so look for deciduous trees that have great surface roots or offer opportunities to improve the root base by grafting or layering. It's much more difficult to improve faulty roots on conifers, so look for specimens that have good roots to start with. Good surface roots are less important for collected junipers as it's common for them to grow in limited spaces that prevent more attractive root structures from developing.

Cork oak with aged bark and an interesting trunk line

Japanese maple with a graceful transition between the trunk and the surface roots

Evaluating Branches

Whether your goal is to create great branch density or delicate branching suitable for a bonsai with a slender trunk, the task of developing branches is a lot easier when there are lots of good branches to choose from. When evaluating branches, the following questions can be helpful.

- Are there enough branches to create a full silhouette?
- Do branches emerge along the length of the trunk, or do they grow only from a few points?
- Are the branches in scale with the trunk?
- Is the trunk movement reflected in the movement of the branches?
- Do the branches look old, or do they look relatively young compared to the trunk?

Standards for deciduous branches are typically higher than they are for branches on conifers or broadleaf evergreens for two reasons. First, because deciduous trees lose their leaves in winter, the branches are on display for a good part of the year. And because it's relatively easy to remove faulty branches on deciduous bonsai and regrow or graft new ones, there can be an expectation that "fixable" problems will be addressed before displaying a tree.

Hinoki

Hornbeam

Evaluating the Whole Tree

Just as you can evaluate a tree's parts, you can ask similar questions about it as a whole.

- Does the tree have special features, such as outstanding deadwood or unusual bark, that make it unique?
- Does the tree have the potential for you to create a novel design or apply a creative solution to its styling challenges?
- Does the tree have presence due to its great age or size?

While it's nice to find trees that check all the boxes, you're more likely to find trees with good and bad points. One reason for this is that it's relatively easy to produce trees with flaws, such as scars, awkward taper, or uneven roots.

When possible, look for trees with minor flaws or flaws that can be mitigated. For example, when considering trees that have prominent scars or wounds on the trunk, ask the following:

- Is it likely that the wound will callus over? Callus tissue forms much easier on some species (such as Japanese maple) than others (such as ginkgo or Yaupon holly).
- Can you conceal the scar with branches or by selecting a new front for the tree?
- If there's no way to conceal a scar, can you transform it into an attractive deadwood feature?

The more you can come up with creative solutions for problems such as scars on the trunk, the more material you can find that will be suitable for development as bonsai.

Satsuki azalea with an interesting trunk, good branch structure, and prominent surface roots

STRATEGIES FOR DEVELOPING A QUALITY COLLECTION

Like so many things worth doing, building a quality collection takes time and effort. Here are some tips for navigating the process.

Growing Your Collection to Support Continued Learning

Investing in education, especially when you're new to bonsai, can pay dividends down the road. You can do this by reading, studying, practicing on your own, or enrolling in online or in-person classes. Taking these steps will help you learn to identify trees with good potential, restore sick trees to health, and perform advanced techniques so you can get the most out of the trees you're working with.

Your goals for learning can also inform your collection-building strategy. One way to do this is to consider what you can learn from a given tree. For example, if you're just getting started with bonsai, you can get a sense of what you can learn from a tree with questions like these:

- Can I keep a tree alive in my garden for more than one year?
- Can I increase a tree's health or vigor?
- Can I successfully repot a tree?

These questions can be answered by acquiring nursery stock or young pre-bonsai (which is why this kind of material is so often recommended for beginners). If, however, you have the basics covered and are ready to improve your skills, you'll need to work with trees that have a substantial trunk (something older than a rooted cutting) and a good number of branches that can help you answer these questions:

- Can I identify an appropriate style for the tree?
- Can I find a good front for the tree?
- Can I do an initial styling?
- Can I create branch pads?
- Can I learn basic wiring techniques?

To learn more advanced skills, look for larger or more complex material that can offer you the opportunity to address the following questions:

- Can I bend heavy branches or trunks?
- Can I create deadwood features?
- Can I learn to graft?

Acquiring older trees or unhealthy bonsai can help you answer these questions:

- Can I help a tree regain vigor?
- Can I restyle a tree that's been in training for a long time?

And when you work with bonsai that already have an attractive shape, you'll face a different set of questions:

- Can I manage vigor?
- Can I refine the tree's design?
- Can I learn to maintain the silhouette?
- Can I prepare a tree for exhibition?

Once you know what you can learn from a tree, you can search for projects that support your learning goals. If you want to practice initial stylings, look for trees with good trunks and lots of branches but little overall shape. If you want to learn about grafting, look for specimens of the species you want to graft that will force you to practice grafting to improve the branch structure.

You can use the same criteria to avoid taking on projects that may not be a good fit for your collection.

- If you don't like wiring, try growing more deciduous bonsai.
- If you don't like scissor work, grow fewer deciduous trees.
- If you don't like pinching new growth every few weeks, avoid fast-growing species such as coast redwood.
- If you like conifers but don't like wiring, try slow-growing species such as ponderosa pine.
- If you don't like pulling needles, don't grow black or red pines.

The more you enjoy working on your trees, the more likely you are to do the work your trees need to improve.

Give Focus to Your Collection

Narrowing the focus of your collection can be a good way to determine whether or not a given tree is a good fit. You can focus your collection based on the following:

- Size (shohin, medium, large)
- Foliage type (conifer, deciduous, broadleaf evergreen)
- Species (such as Japanese maple cultivars, Satsuki azaleas)
- Style (such as literati bonsai or forest plantings)
- Geography (such as alpine conifers native to the Rocky Mountains)
- Quality (trees acceptable for display at regional exhibits)
- Stage of development (show trees, long-term projects with good potential)

It's okay to select more than one focus for your collection. The main idea is that if you can determine which trees you most enjoy having in the garden, you may find it easier to acquire additional trees that provide a similar level of enjoyment. Growing or collecting "one of everything" can work, but different sizes and species can have differing water, light, and climate requirements that can be challenging to provide. On the other hand, growing trees in batches can do a lot to accelerate your learning. When you grow more than one of a given species, you have more opportunities to try out different approaches and learn which work best for you.

The Most Important Thing You Can Do to Improve Your Collection

By far, the number one thing you can do to improve the quality of your collection is to think carefully about the trees you bring into your garden.

If you've been growing bonsai for a while, you've learned to see the potential in trees, and you've learned the techniques required to realize this potential. However, maintaining too large a collection can prevent you from improving the trees that mean the most to you. Depending on the species you grow, creating or maintaining as few as fifteen show-quality trees can be a lot of work.

In general, the number of quality trees in your collection is a reflection of how much time you have to spend on them. If you're not happy with the rate at which your trees are improving, you might try reducing the size of your collection or finding a professional to help you keep up with the work.

I'll admit that limiting the number of trees in a collection is easier said than done. Some of the trees in my collection don't get the attention they deserve simply because I have too many other trees to work on and I can't get to them all. For years, I thought I could maintain these trees without causing them to decline, but I've learned repeatedly that it's nearly impossible for trees to improve, or even maintain their current state, without giving them adequate attention.

If you already have more trees than you can manage and would like to cull your collection, you can start by using the criteria in the previous sections to evaluate which of your trees have the most potential to help you achieve your goals. For the others, you can take advantage of local bonsai clubs or online communities that can help you find new homes for any trees you no longer need.

BONSAI SHOPPING

PLACES TO SHOP FOR BONSAI

— Online bonsai vendors (inventory can be variable, particularly for smaller shops, so investing time in the search and checking sites regularly can help you find great trees)

— Retail bonsai nurseries (good opportunities for hidden discoveries, but requires visiting in person, which may not be convenient depending on how far you have to travel)

— Vendor areas at bonsai shows (great material may be available, but there are only so many bonsai shows in a given region)

Trees also change hands at private sales, auctions, and special events. The stronger your ties to your local bonsai community, the more of these events you'll learn about. Helping in a friend's garden, volunteering at a public bonsai collection, or offering to help at a bonsai nursery can open doors to these additional sources of material. In short, the more often you're around quality trees when they become available, the more likely you are to add one to your collection.

SHOPPING TIPS

— Keep in mind that for most bonsai, the bulk of the value is in the trunk. More often than not, it's easier to create a great tree by starting with a good trunk that has flawed branches than with a flawed trunk that has attractive branches.

— Try to distinguish between problems that are easy to fix (such as trees that need restyling) and problems that aren't easy to fix (such as large, unsightly scars on the trunk or awkward transitions of taper).

— Look for rescues. If your horticultural skills are good, shop for trees with great potential that may need a few years of recuperation before you can begin styling work.

— When shopping, don't be afraid to go home with nothing—not bringing home another tree can be a sign of success when your goal is to make your collection the best it can be.

TIPS FOR CREATING A QUALITY COLLECTION ON A LIMITED BUDGET

— Buy trees in earlier stages of development (the same money goes further when you do the work).

— Grow trees from seed, cutting, or air layer (this approach requires a broad set of skills and takes a long time but can be incredibly rewarding).

— Look where other people aren't looking. Overgrown material at retail nurseries and discarded trees and shrubs from garden renovations can provide great starting points.

EXAMPLES: SELECTING BONSAI BASED ON THEIR POTENTIAL

Blue atlas cedar

Shimpaku juniper

Black pine

Trident maple

When you're shopping for quality starting points, the idea is to focus more on what a tree could become rather than what it looks like now. Here's a look at a number of trees that could be good additions to a collection as learning trees or, in other cases, as show trees.

Good Trees for Building Your Skills

If your goal is to improve your skills, look for trees that give you an opportunity to practice the skills you're looking to improve. These example bonsai are good for practicing wiring and styling but offer little opportunity for making big transformations within a short timeframe.

BLUE ATLAS CEDAR

- The trunk has little movement or taper and has yet to develop craggy bark.
- The branch structure is poor (which gives the tree a two-dimensional appearance) and the upper branches are large in relation to the trunk.
- The best opportunities for quickly improving the tree are to wire the branches and create attractive branch pads that can fill in over time.

SHIMPAKU JUNIPER

- The trunk movement is simple (there is a single curve with no twist).
- The branch structure is poor, as most branches emerge from the same point along the trunk.
- Because the branches are slender enough to bend, this tree is a good candidate for styling and developing branch pads, which can make a big improvement in as little as two to three years.

BLACK PINE

- The trunk movement is repetitive (the curves are the same size) and the tree has little taper (the first change in trunk size is near the top of the tree).
- The branch structure is poor, with too few branches and too much space between them.
- Although it won't address the flaws in the trunk, grafting additional branches can improve the branch structure and in time create a full silhouette.

TRIDENT MAPLE

- This tree is young. The trunk has simple movement and the branches have yet to develop any character.
- The best opportunity for improving this tree would be to grow it over a long period of time (more than ten years) to create a more compelling trunk with better taper and movement.
- While it's possible to learn basic techniques for developing trident maples with this tree, starting with an older tree offers more opportunities to try out techniques and learn the ins and outs of the species.

Pre-Bonsai with Potential

These junipers are good examples of pre-bonsai that have been cultivated for development as bonsai with a focus on growing the trunk but not the branches. The grower did a great job creating these trees by giving the trunks interesting movement and taper. Although there is no deadwood, the multiple trunk lines offer ample opportunities for creating jin and shari.

Trees at this stage of development are great for learning front selection, grafting, branch pad creation, and carving. Both trees have good enough trunks to be shown at local or regional exhibits in as little as three to five years. (For an example of what trees like this can look like after a few years of training, see the semi-cascade juniper created by the same grower on page 27.)

Shimpaku juniper

Shimpaku juniper

Shohin Shimpaku Junipers

The shimpaku junipers shown here are relatively young (about ten to fifteen years old) and small (both are considered shohin bonsai because they are less than 8 inches tall). At a glance, both have movement in their trunks and deadwood features, two of the most important characteristics for juniper bonsai. The difference is in the character of the trunk movement. The curved trunk of the informal upright juniper is simple, while the trunk of the semi-cascade juniper has more twists and turns.

In terms of the branches, the informal upright juniper has relatively few branches, and they emerge from just two points along the trunk. The semi-cascade juniper has branches along the length of the trunk, which increases the opportunities for creating an attractive design. (You can see the creation of the semi-cascade juniper on pages 56–59 in *The Little Book of Bonsai*.) Although both trees are good for practicing basic bonsai techniques, the semi-cascade juniper has the greater potential to become a quality bonsai.

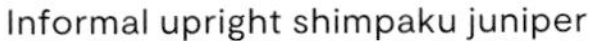

Informal upright shimpaku juniper

Semi-cascade shimpaku juniper

Procumbens and Shimpaku Junipers

The junipers below are both medium-sized bonsai (less than 18 inches tall) that feature shari along the trunk. The big differences between the two are in the trunks, the branches, and the surface roots.

The procumbens juniper has far more interesting movement than the shimpaku, which, in contrast, has relatively simple movement. This characteristic alone makes the procumbens a better candidate for developing into a quality bonsai.

Although surface roots aren't generally an important characteristic of juniper bonsai, they can demonstrate the tree's connection to the soil. The shimpaku juniper roots are somewhat exposed and would look better (less precarious) if they were planted several inches deeper in the pot.

As for the branches, it's clear that the procumbens juniper's branches have recently been wired, while the shimpaku juniper's branches haven't. In general, whether or not a tree has been wired has relatively little bearing on its value, as it doesn't take much effort to apply wire. What does make a difference is the quality and distribution of the branches. The procumbens has compact branches that can easily be arranged to create an attractive silhouette for the tree. The shimpaku has fewer branches that are less well-distributed along the trunk.

Shimpaku juniper

Procumbens juniper

ANDREW ROBSON

Seeing great bonsai can have a lasting effect on the viewer. In some cases, it can change the viewer's life. That was the case for bonsai professional Andrew Robson. In 2015, he visited the Artisans Cup, a bonsai event that featured spectacular trees from around the United States. After experiencing such great trees in person, Andrew realized that none of his bonsai would be accepted into such a show. At that point, he set a goal of creating a garden in which every tree was worthy of display in a top-tier exhibit. Less than ten years later, Andrew has made great progress toward this goal.

Although not every tree in his collection is ready to be shown, all are well on their way. Here's how he did it. The first step was changing his mindset. When shopping for trees, he limited his options to those that had the potential to become great. The next step was disregarding the time it would take to develop them. It didn't matter to Andrew whether it would take five, ten, or even fifteen or more years. In some cases, he'd stand before a tree and ask himself, "In time, could this tree get into the best show in the country?" If the answer was yes, he'd go for it.

The final step was ensuring that all the trees in his garden received adequate attention to develop into great bonsai. This last step requires diligence, and Andrew has a few tips to make the process easier:

— Narrowing focus in terms of the species or sizes of trees in your collection reduces the mental load required to maintain different species at a high level. Andrew, for example, focuses his collection on large deciduous bonsai.

— Maintain no more trees in your collection than you have time for. When time is constrained, consider reducing the size of your collection to ensure the trees you care about get the attention they deserve.

— Growing trees that you love will encourage you to do what's best for them. Focusing your collection on species that grow well where you live and that provide you with opportunities to do the kind of work you enjoy will make spending time in the garden a pleasant activity year-round.

CHAPTER 2

Bonsai Design

Studying bonsai design is a great way to help you understand the options you have for styling your trees. The goal for this chapter is to provide key concepts to help you navigate bonsai design decisions. We'll start by looking at the stories your trees can tell.

DESIGN CUES: SELECTING MODELS FOR YOUR TREES

One way to approach bonsai styling is to think about the "stories" your trees tell. Depending on how it is styled, a bonsai can suggest the environment in which it developed, the length of time it's lived, and the traits inherent to the species. Bonsai can convey these characteristics with varying degrees of realism or with great artistic license. As the artist responsible for your trees' development, the choice of which characteristics your trees convey and how they do it is up to you.

Telling a Story of Place

All trees tell a story of place. As a tree grows, it takes on characteristics that reflect the environment in which it matures. A mild environment, for example, may produce trees with relatively straight trunks and full, rounded silhouettes. Harsh environments are more likely to produce twisted forms with lots of deadwood.

These same ideas influence bonsai design. In some cases, these characteristics inform the basic style of the tree. Wind-influenced designs, for example, can tell a story of trees growing in environments defined by prevailing winds. Cascade-style bonsai can tell a story of life on a mountain slope. If you imagine what kind of environment might have produced the tree you're working on, you can play up these features to tell a more compelling story of the place, real or imagined, that your tree suggests.

Sierra juniper

Conveying a Tree's Age

Imagine how a tree looks at different stages of its life. You might envision a seedling, a young tree, a full-sized tree, an aged tree, and a declining tree. Although it's not hard to find examples of bonsai representing younger trees, most are styled as idealized versions of trees in later stages of life.

There is a strong convention to style bonsai to convey great age, and to a large degree, the work of training bonsai focuses on accelerating the production of characteristics that nature produces over time. Creating jin and shari, for example, approximates the dieback along the trunk and branches that we might expect to see in a tree that has grown old in a harsh environment.

Another way to suggest great age is to create a canopy with an asymmetrical form. Trees become more distinct as they age and are affected by their environment. Over time, branches can break in storms, and limited light or extreme wind can influence branch length and density. The older the tree and the harsher the environment, the less likely the tree will maintain a symmetrical silhouette.

To make your tree look older, start by determining which age-related features it already has so you can highlight or exaggerate them effectively.

Young pine, mature pine, very old juniper

Telling a Story of Scale

Compared with mature trees growing in nature, bonsai are relatively small. That said, not all bonsai are miniature versions of larger trees. Many collected trees, such as the shimpaku juniper below, are styled at actual size. These bonsai typically show their age in the texture of their bark or deadwood, features that are best appreciated up close.

Other species, including coast redwood, cryptomeria (shown here), and zelkova, for example, are often styled to look like miniature versions of full-size trees viewed from a distance.

If you want to make your bonsai look bigger or smaller, you can do that by modifying the size of the branch pads. For example, styling a tree so it has a greater number of small branch pads can make it look bigger. Styling a tree with fewer branch pads that are larger can make the tree look smaller. And if you get the balance right, it can be hard for a viewer to determine the size of the tree when looking at a photograph of it.

Shimpaku juniper with intricate deadwood features

Cryptomeria styled to look like a large tree viewed from a distance

Species-Specific Design Cues

Each species has a unique set of operating instructions that determine how it responds to its growing conditions. These traits help make a pine look like a pine and a maple look like a maple.

Sierra junipers, for example, produce straight trunks when they grow at lower elevations and develop sinuous, twisting forms known as *krummholz* (German for "crooked wood") when they grow in harsh environments. When we see Sierra juniper bonsai that have twisted forms, we can appreciate that these trees "look like Sierra junipers" because they exhibit characteristics associated with junipers that grow at higher elevations.

Evergreen oaks are known for producing long, undulating branches that meander as they extend from the trunk, in part because oak leaves grow in a repeating spiral pattern. When styling oak bonsai, it's common to create branches with interesting movement, as this can evoke the feeling of oaks growing in nature. Of course, not every tree fits this pattern. Species that are naturally shrubs, such as boxwood, for example, are often styled as oaks because they lack a characteristic form of their own.

The more familiar you become with the traits of the species you work with, the easier it will be to express your trees' natural beauty by bringing out their best traits.

Design Cues Based on the Degree of Realism

The next time you work on one of your trees, ask yourself if you can imagine a similar tree growing in nature with the same proportions. If so, the tree likely exhibits characteristics common to *naturalistic bonsai*. If that's not the case, you may be working with more *abstract*, *idealized*, or *expressionistic bonsai*.

Wind-influenced designs provide an example of expressionistic approaches to styling. The effect, however, can appear overdone if the resulting tree looks like it's being blown by a gale-force wind.

Bonsai that feature large trunks with dramatic taper, gentle movement, and prominent surface roots often look like idealized versions of what we might expect to find in a mature tree. Many Japanese black pine and trident maple bonsai are trained in this fashion.

These are just a few examples of different approaches you can take with your trees. As you select models to fashion your bonsai after, whether they're inspired by your trees' age, size, or the environment in which they grow, strive to highlight your trees' best features to help them convey the most compelling stories possible.

Trident maple with exaggerated or idealized features

DESIGN TIPS FOR STYLING YOUR TREES

The remainder of this chapter focuses on practical styling tips to help you make design decisions for your trees. We'll start by describing a process for finding the front of the tree and then look at how to create a branch structure that complements the trunk and enhances the tree's flow.

Finding the Front

To find the front of the tree, look at it from all sides and test out different planting angles to determine which you like best. While you do this, consider:

1. The trunk

   Which side shows the most interesting movement?

   Which side has the most attractive taper?

   Which side is the most graceful/powerful?

2. The surface roots

   Which side has the most attractive roots?

   Which side has the most attractive transition from the lower trunk to the roots?

   What is the most natural planting angle for the roots?

3. The branches

   Which side offers the best distribution of branches from left to right and from front to back?

   Which side doesn't have so many branches that they block your view of the trunk?

   Which side shows off particularly attractive primary branches?

4. How you can highlight special features

   Which side has the best bark?

   Which side has the most aged or interesting deadwood features?

5. How you can downplay flaws

   Which side has the fewest scars on the trunk?

   Which side hides awkward taper best?

There's no formula for navigating the inevitable trade-offs between options, but there are a few starting points.

— Focus on making the trunk look as good as possible and adjust from there based on other considerations. Much of a tree's age, character, and beauty are found in the trunk, so presenting the most attractive view of it will help you get the most out of the tree's potential.

— Orient the trunk so the top of the tree is in line with, or slightly in front of, the base of the trunk. It's unconventional for bonsai to lean away from the viewer.

— Test out different fronts ahead of repotting time. One approach is to look for an obvious front, an alternative front, and an off-the-wall or creative option for the front and then compare the good and bad points of each. There is more than one good option for the Rocky Mountain juniper shown on the next page. Do you have a favorite?

Rocky Mountain juniper—candidate front 1

Candidate front 2

Candidate front 3

Candidate front 4

Designing the Branches and Canopy

The shape and character of a tree's trunk often provide cues for how to arrange the branches. For example, if the trunk is thoroughly twisty, you may find that it doesn't make sense for the branches to be perfectly straight. Likewise, if the trunk is relatively straight, you may decide that it doesn't make sense for every branch on the tree to feature dramatic curves. How you interpret design cues in the trunk and complement them with a branch design makes your tree uniquely yours.

If you study the relationship between trunks and branches, you may start to notice some common patterns.

— Trees with wider trunks tend to have longer branches. Trees with slender trunks tend to have shorter branches. (It would be strange, for example, for a coast redwood to have a slender trunk but wide branches, as this wouldn't be a stable arrangement.)

— Trees with wider trunks tend to have full silhouettes (less space between branches). Trees with slender trunks tend to have less full silhouettes (more space between branches).

— For conifers, primary branches are more likely to grow upward on young trees and downward on old trees. For deciduous and broadleaf evergreen trees, primary branches can grow upward or downward.

— Younger (or smaller) trees tend to have more simple branch development. Older (or larger) trees tend to have more complex branch development.

Tiger bark ficus—the broad trunk suggests a canopy that's wider than it is tall

Another pattern that can inform your styling relates to branch angle consistency. On many trees, the primary branches will emerge from the trunk at roughly consistent angles. What's less common is for branches to grow at seemingly arbitrary angles, as the same forces that influence how branches grow tend to influence all of the branches, not just some of them.

When studying branch angles, take note of how high (or low) the lowest branch emerges from the trunk. On junipers, for example, it's common for the lowest branch to emerge halfway up the trunk or higher. In general, the steeper the angles at which the primary branches grow, the higher along the trunk these branches will emerge.

Juniper with steep branch angles

Creating Coarse to Fine Structure

The principle of "coarse to fine" in bonsai is the idea that large structural elements of a tree (such as the trunk or a branch) become smaller toward the periphery of the tree. This idea provides a framework for a lot of the patterns found in trees that grow in nature.

- The lower branches are thicker than the upper branches.
- The base of a branch is thicker than the extremity of a branch.
- The internodes at the base of a branch are longer than the internodes at the extremity of a branch.
- The lowest part of the trunk is thicker than the top of the trunk.
- The lower branch pads are larger than the upper branch pads.
- The gaps between large lower branch pads are bigger than the gaps between smaller branch pads on the upper branches.

Instead of interpreting coarse to fine as a rule for training bonsai, you can apply it as a guide that can simplify some design decisions. When deciding which branches to remove in a crowded area, for example, you might opt to keep larger branches lower on the trunk and keep smaller branches near the apex.

Determining Flow

When viewed from the front, nearly all bonsai "flow" from one side to the other (from left to right or right to left). Here's an example of an atlas cedar that flows to the right.

Several features can indicate flow in bonsai:

- The line of the trunk
- The direction of the key branch
- The shape of the silhouette (particularly near the apex)

If you start at the base of the tree and trace the line of the trunk, the primary direction in which it moves indicates the flow of the trunk.

The key branch is typically the most prominent branch on the tree. If it grows on the right side of the tree, it indicates flow to the right.

The sides of the upper canopy can indicate flow based on which of the two sides is steeper.

In the case of this cedar, the trunk, the key branch, and the silhouette all "point" or flow to the right.

Atlas cedar

Although it's common for these three indicators to flow the same way, it's perfectly acceptable for different parts of the tree to flow in different directions. For example, the trunk of the cork bark elm shown here flows to the left and the key branch flows to the right. (The upper silhouette is close to symmetrical but flows slightly to the right.) Overall this tree flows to the left as the degree to which the trunk leans left is greater than the degree to which the canopy and key branch indicate movement to the right.

When the trunk slants one way, it's common for the silhouette and/or the key branch to flow the opposite way as this can create a balanced composition (one that doesn't look tippy).

To exaggerate the flow in a given direction, lengthen the branches in the direction of the flow and shorten the branches on the other side of the tree. For example, to exaggerate the flow to the right, lengthen the branches on the right side and shorten the branches on the left.

Cork bark elm

Determining whether a tree flows left or right can inform decisions about which branches to keep, how long or short to make branches, or what shape to give the upper canopy. If something about a tree feels off or unbalanced to you, determine the flow of the tree and look for opportunities to improve it.

As an experiment, evaluate the flow of the trees pictured in this book. Some trees have clearly defined flow, while others have yet to develop strong flow one way or the other. If you see a tree with ambiguous flow, ask yourself if you can think of a few changes that can clarify the flow and improve the design.

Western juniper

For Further Study—A Bonsai Design Exercise

The best way to refine your approach to design is to study beautiful trees. When possible, do this in person at exhibitions or in gardens you admire. When you get up close to a tree, you can follow the trunk line in three dimensions, evaluate primary branch angles and apex construction in detail, and appreciate special features such as deadwood or old, flaky bark in a way that's hard to appreciate in two dimensions.

Studying images of trees can also provide great opportunities for learning. You can compare the size and shape of the trunk to the size of the silhouette, study the angles at which branches extend from the trunk, and consider options for improving a tree's flow.

For more active study, trace the trunks of several bonsai you admire or draw imaginary trunk outlines. Next, draw the primary branches and indicate the overall silhouette of the tree. When you're done, compare your results with the original photo of the tree.

If you don't like the result, try again. The more you practice identifying good locations for primary branches or creating canopies that make sense for the trunk, the easier it will be to create successful designs for your trees.

Korean hornbeam

Trunk outline

A rounded canopy with branches that grow downward—
a common pattern for conifers

A pointed canopy with branches that grow upward—
a common pattern for deciduous bonsai

A pointed canopy with branches that grow downward—
a common pattern for conifers

A rounded canopy with branches that grow upward—
a common pattern for deciduous bonsai

CHAPTER 3

Applying the Right Techniques at the Right Time

One of the easiest things you can do to speed development of your trees is to make sure you're applying the right techniques to the right trees at the right time. This chapter will help you do this by identifying appropriate techniques for different stages of development and noting the best time of year to perform common bonsai training techniques.

IDENTIFYING DEVELOPMENT GOALS

In the previous two chapters, we looked at criteria for selecting trees that would be a good fit for your collection and starting points for determining how you want your trees to look. Armed with this information, you can take the next step of identifying development goals for your trees.

Unlike a discrete task you might perform on a tree (such as removing a branch or finding a new front), a development goal is something that you typically work on over time. Development goals can include things like increasing the size of the trunk, defining the silhouette, or refining a mature tree by improving the branch structure and density. This is when you can put to use everything you've learned about bonsai style to help you envision the future of what your trees can become. Once you have identified your goals for a tree, you can determine the scope and sequence of the work.

To help you get an idea of which goals make sense at different stages of development, the following section provides an overview of how to grow trees from scratch.

Overview of Bonsai Development

When you grow bonsai from seed, cutting, air layer, or nursery stock, the process differs greatly from projects that begin with a mature trunk (such as the trees you dig from the mountains or from a garden). After growing bonsai from seed and cutting for more than three decades, I've learned that working with young trees is a great way to learn what's possible with a species, as it's easier (and less costly) to experiment with young material than with mature specimens. I've also found that the more familiar you are with well-developed, mature bonsai, the easier it is to create trees from scratch that emulate the best characteristics older bonsai have to offer.

To help you identify which techniques make sense for different stages of development, we can divide the process into three basic steps that focus on the roots, the trunk, and the branches.

STEP ONE: ESTABLISH AN ATTRACTIVE ROOT BASE

The focus for the first few years of a tree's life is to create a starting point that is good enough for you to invest your time over the next five to ten (or more) years.

- Select seeds (from parent trees with good traits), cuttings, or young trees with good characteristics for the species (small leaves and internodes for deciduous species, good bark and needles for pines, and so on).
- Establish attractive surface roots by making incremental improvements every time you repot (arrange supple roots before they get too big to move, shorten large roots to encourage finer root growth, remove roots that emerge above or below the main root base).

Although this initial stage is brief, it sets the foundation for all future development. If, for example, you plant a young tree with faulty roots in the ground and quickly increase the size of the trunk, the odds are slim that the roots will automatically look better once the trunk reaches the desired size.

Once you're happy with the material you've selected and the roots are in good shape, you can shift your focus to building the trunk.

STEP TWO: BUILD THE TRUNK

Building the trunk requires more time than establishing the root base, as building a quality trunk generally takes a minimum of five to ten years, depending on the size and style you're aiming for. When focusing on trunk development:

- Create an attractive shape for the trunk (typically by pruning and wiring).
- Create taper in the trunk that's appropriate for the style you select for your tree.
- Increase the size of the trunk until it reaches the desired thickness.
- Develop characteristics that add character to the trunk and the appearance of age (such as fissured deadwood features on junipers).

The most effective way to thicken the trunk quickly is to use sacrifice branches. A sacrifice branch can be any branch used to thicken the trunk (or part of a larger branch) that you remove when the trunk (or part of a larger branch) reaches the desired size. To encourage rapid trunk development, you can plant young trees in the ground or use successively larger containers when you repot.

Because this step takes longer than the others, it's a common tendency to want to cut trunk development time short. Doing so results in trees that lack character or attractive transitions of taper. To avoid this, apply the criteria for evaluating the trunk in Chapter 1 and see if there's anything else you can do to improve the trunk before removing sacrifice branches and focusing on branch development.

STEP THREE: DEVELOP THE BRANCHES

When the trunk reaches the desired size, shape, and character, your focus can shift to branch development.

- Establish primary and secondary branches to create a basic silhouette.
- Use species-specific techniques such as defoliation or decandling to reduce internode length and improve branch density.

During this step, the work transitions into what we might think of as "basic bonsai development techniques"—the bulk of the training techniques featured in this book. Instead of using successively larger containers as you did when developing the root base and trunk, you can begin using smaller containers to keep new growth in check. And instead of encouraging vigorous growth, you can reduce vigor to produce more refined branches.

Here's an example of what the process looks like from start to finish. The black pines shown here are roughly three, five, seven, eight, ten, and thirty years old. The youngest tree is in the first stage of development where the goal is to create a good root base. The second and third trees are in the second stage where the goal is to build the trunk. The fourth and fifth trees are in the third stage of development in which the goal is to create an attractive branch structure and silhouette. The sixth tree is in the refinement stage. It has an established design but needs additional work so it can continue to improve as it ages.

Japanese black pines at different stages of development

ALIGNING DEVELOPMENT TECHNIQUES

In general, bonsai work in early stages of development requires rapid growth. In later stages of development, the tree's growth slows down as your focus shifts from increasing trunk size to increasing branch quality and density. Here's where aligning techniques with stages of development comes into play. If your goal is to thicken the trunk, techniques that reduce vigor such as decandling or defoliation can become counterproductive.

A related idea is that the techniques you apply to your trees need to change as your trees change. Once a tree reaches a certain level of branch density, for example, you'll need to shift the focus from building density to thinning to ensure that each branch receives enough light to stay healthy. When you can successfully align training techniques with your development goals, you can accelerate development and make your trees look better every time you work on them.

PETER TEA

Peter Tea's garden is known as a place where trees take shape over time. Instead of quickly styling a tree and moving on to the next one, Peter chooses trees with great potential and then invests the time it takes to improve them over a period of years, not hours. This isn't surprising considering that Peter apprenticed at Aichi-en, the Nagoya-based garden of bonsai professional Junichiro Tanaka.

I spent six weeks working with Peter and Tanaka at Aichi-en and saw the work they did firsthand. More so than I've seen at other bonsai gardens in Japan, the work at Aichi-en focused on trees at different stages of development. Some of these bonsai have been in the family for generations, including an ume (Japanese flowering plum) started from seed by Tanaka's great-grandfather in 1896.

When Peter completed his apprenticeship and returned home to California, he got to work on his own long-term projects. His guiding principle for building trees is what he refers to as the "inside out concept." The basic idea is to create bonsai incrementally starting with the core of the tree, the base of the trunk, and then building outward, one increment at a time, as branches divide.

"We start at step one and make sure that we're satisfied with the trunk size and shape before moving to step two," Peter says. Following this sequence can clarify what work happens when as there's no need to worry about the final silhouette when you're still building the primary branches.

Peter conveys these ideas through a diagram he developed "to help students look deeper into the tree to see what steps have already been completed and what needs to happen next." The numbers in the diagram refer to the sequence in which each area of the tree is developed.

SCHEDULING THE WORK: A BONSAI CALENDAR

After determining which kinds of techniques to apply to your trees, your next step is to determine when to apply them. The following calendar organizes common bonsai tasks by time of year. There's no mention of months or dates as the specifics will depend on where you live, how your trees are growing, and your local weather.

Because not all work fits neatly onto a calendar, here are some tips related to annual work cycles:

- Junipers are relatively forgiving about when they are pruned or wired. They are commonly styled in summer, fall, and winter, but minor work can be done at any time.
- Although you can create jin and shari year-round, it's common to do the work in spring when trees are growing quickly, as the bark separates easily at this time of year due to increased sap flow.
- Some species such as coast live oak follow much less precise growth cycles. When conditions are favorable, a live oak can produce two to three flushes of new growth a year. When a live oak is weak, it can go one or two years without producing any new growth at all. If you work with species that follow less precise schedules, follow the tree's lead, and don't hesitate to postpone work when there's nothing to do.

Managing Stress on Trees

One of the hardest things to manage when sequencing bonsai work is considering how stressful a given technique can be for a tree. A major repot, for example, can slow a tree down for a full year, especially if it wasn't healthy to begin with. After a big repot, it may make sense to postpone stressful techniques such as defoliating, decandling, or bending large branches until the tree has fully recovered.

Over time, experience will help you determine when you can or can't work on a tree. In the meantime, whenever you have stressful work planned for a tree, consider what it has been through in the previous twelve months and look for signs that it's growing well before getting started.

SEASONAL CARE CALENDAR

LATE WINTER/EARLY SPRING
(when you first see signs of new growth)

— Repot conifers and deciduous bonsai (non-tropical species).

— Graft conifers and deciduous species.

EARLY SPRING
(as new buds emerge)

— Pinch initial elongating buds on maple, beech, and hornbeam.

— Pinch buds on maple, elm, zelkova, and coast redwood throughout the growing season.

MID-SPRING
(before new growth hardens off)

— Pinch candles on black or red pine after candles elongate (see page 80).

— Pinch spring growth on cedar, spruce, and high mountain pines (ponderosa pine, white pine, mugo pine) before it hardens off.

— Thin dense deciduous bonsai when the exterior foliage shades out interior branches.

— Start air layers as spring growth starts to harden off.

LATE SPRING/EARLY MIDSUMMER
(when the spring flush hardens off)

— Prune all deciduous and broadleaf evergreen bonsai.

— Prune hinoki cypress, juniper, cedar, spruce, and redwood.

— Decandle red and black pines (see page 80), thin needles, prune, and apply wire.

— Fully or partially defoliate (see page 80) silverberry and deciduous species such as Japanese maple, trident maple, zelkova, and hornbeam.

— Thread graft defoliated deciduous species such as trident maple or Japanese maple.

— Cut large leaves (see page 82) on silverberry and on deciduous species such as stewartia, Japanese maple, and Chinese quince.

— Repot chojubai (dwarf flowering quince) that weren't repotted in spring.

— Prune, wire, and repot tropical species including ficus, jade, and bougainvillea when temperatures are warm.

MID- TO LATE SUMMER
(when trees have produced new growth after spring pruning/when foliage on high mountain pines has hardened off)

— Defoliate or cut leaves on silverberry and maple.

— Prune maple, elm, zelkova, and olive.

— Prune and wire junipers and high mountain pines.

FALL
(when deciduous leaves have mostly turned color/when summer growth is mature on decandled pines)

— Remove old leaves on deciduous species.

— Pull needles on black and red pines.

— Prune and wire conifers, deciduous, and broadleaf evergreen species.

WINTER
(when trees are dormant)

— Prune and wire conifers, deciduous, and broadleaf evergreen species where winter is mild and trees are protected from hard freezes.

CHAPTER 4

Pruning and Wiring Basics

A large portion of bonsai work focuses on two basic tasks: removing growth (by pruning) and redirecting it (by wiring). The goal for this chapter is to introduce the fundamental pruning and wiring techniques upon which more advanced techniques are based to help you create the designs you have in mind for your trees.

PRUNING BASICS

Dotted line indicating the desired silhouette

We prune bonsai for a variety of reasons: to improve branch structure, to stimulate new buds, and to give trees shape and maintain their size. These reasons give us a variety of criteria we can use to determine when, where, and how much to prune.

One set of criteria applies to trees as a whole. To start with a simple example, when long shoots extend beyond the desired silhouette, it's easy to see where to prune to maintain the size and shape of the tree.

You can also consider how much foliage you can remove from a tree at a particular time. For a mature black pine, for example, it's safe to remove up to 50 percent of the foliage in fall.

Striking a balance between removing too much and not enough requires practice. Removing too much foliage at any given time can weaken a tree and lead to branch dieback. Not pruning enough can cause branches to become too thick, particularly near the top of the tree, and cause lower branches to become weak when they don't receive enough light.

In general, the idea is to remove as much foliage as you need to achieve your goal, but no more than that to avoid making a tree weak. To help you get a sense of how much work can be done at any given time, Chapters 9 and 10 feature case studies that demonstrate many of the pruning techniques described in this book.

Pruning to Stimulate Budding

When you prune a healthy branch, you can stimulate the production of one or more new buds. For example, if you prune the oak branch shown here at the dotted line, you can expect the bud closest to the cut site to grow.

The strength of the branch and the type of tree you're pruning will determine how many buds open. For example, if you shorten a weak branch in the tree's interior, the cut is not likely to stimulate new growth. If, however, you shorten a long, vigorous shoot growing near the top of the tree, the branch is likely to produce several new shoots near the cut site.

There are two main criteria for determining how much of a branch to remove. The first is to consider branch length in terms of the shape of the tree. If a branch extends beyond the desired silhouette, you can prune back to the silhouette you have in mind for the tree. The second is to consider the internode length. If a branch grows too long before it divides into smaller branches, it can be difficult to create a sense of scale or suggest a larger tree growing in miniature. In these cases, you can prune branches to the desired internode length to create an appropriate branch structure for the tree. This might mean pruning branches back to about one ¼ inch to ½ inch for a small tree or up to 1 inch or more for a larger tree. Over time, these smaller branch segments can fill in the silhouette you have in mind for the tree.

Oak branch

After pruning

You can also make pruning decisions based on the direction in which you want new buds to grow. In general, deciduous species produce buds singly or in pairs. Japanese maples, for example, produce buds in pairs. This is known as an opposite budding pattern. When you prune back to a pair of buds on a Japanese maple, you can expect both to open. Species such as elm and hornbeam have what's known as an alternate budding pattern in which a bud on one side of the branch is followed by one on the other side, back and forth along the length of the branch. When pruning alternate budding species, you can prune to a bud that points in the direction you want the branch to grow. This practice is known as *directional pruning*.

Directional pruning can create beautiful branch structures and simplify the task of wiring as there's less bending to do when branches are already growing in the right direction.

The pattern differs slightly when pruning evergreen conifers such as pine, spruce, or juniper. Evergreen conifers don't produce new buds as easily as deciduous species or broadleaf evergreens, so the best way to maintain branch health is to preserve healthy foliage (including a healthy bud for species like pine, spruce, and cedar) at the end of every branch. If you remove all of the foliage at the end of the branch, the remaining woody portion will likely die back to the nearest branch with healthy foliage.

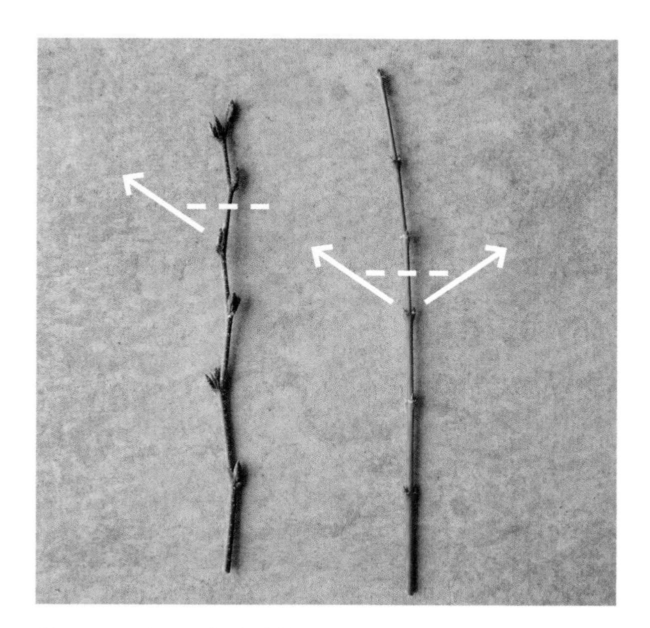

Alternate and opposite budding patterns with arrows indicating the direction in which new buds will grow

Fine twigs on stewartia bonsai

Pruning to Create a Simple Branch Pad

Branch pads are blocks of foliage that are typically built around a primary or secondary branch. You can create branch pads on any kind of bonsai, but they are more common on conifers and broadleaf evergreens than on deciduous bonsai.

Branch pads often feature a central leader (a main branch) that divides into smaller and smaller branches toward the branch tip. Pads are typically arranged so that most of the foliage appears on the top and sides of the pad, as these are the surfaces that receive the most light.

The facing page shows an example of how to create a simple branch pad by pruning.

GOALS FOR PRUNING THIS BRANCH

— Remove the elongating shoots to create a rounded pad shape.

— Remove downward growing shoots.

— Thin foliage in dense areas to create even foliar density.

Pruning the elongating shoots that extend beyond the silhouette of the pad creates a more rounded shape. Thinning the relatively dense foliage facilitates wiring and lets light pass through the pad to any branches below.

Juniper branch before pruning—top

Before pruning—side

After pruning—top

After pruning—side

When to Prune Bonsai

If you review the calendar in Chapter 3, you'll find that pruning is recommended during most times of the year. This suggests that there's some flexibility for when you can work on your trees, particularly for younger specimens. As trees mature, however, the timing becomes more important. A few general principles related to how trees grow can help you determine the best times of year to prune.

For example, at the start of the growing season, trees use stored resources to produce new growth in spring. Trees rely on this new growth to replenish their resources so they can repeat the process the following year. When we prune early in the season, we reduce the tree's capacity to generate resources over the course of the growing season. When we prune later in the year (in fall or winter), trees can take advantage of any new growth produced in spring. With this basic idea in mind, you can begin to think about why it makes sense to prune some trees early in the year (when you want to slow down a vigorous tree) and prune others later in the year (when you want them to gain or maintain vigor).

Here's how this works in practice. In general, if you prune once a year, prune in fall to give trees an opportunity to use spring growth to photosynthesize and build up their resources over summer. If you need to prune twice a year, try pruning after the spring flush hardens off and again in fall. And if you're working with a vigorous species that requires frequent pruning (such as trident maple, zelkova, or silverberry), prune as often as needed during the growing season.

In most cases, a tree will let you know when it's ready to be pruned. If, for example, the tree grows slowly and produces little new foliage, it may be healthy but not vigorous. In this case, there's generally no need to do any pruning. If, on the other hand, it produces lots of new growth, the tree is a good candidate for pruning.

The juniper branches shown here provide an example of healthy foliage that's growing slowly, healthy foliage that's producing strong new growth, and a vigorous new shoot. In general, branches for shoot 1 don't need to be pruned (at most, do light thinning). Shoots 2 and 3 are good candidates for pruning and wiring.

Healthy, strong, and vigorous juniper growth

Timing can also come into play when removing large branches. When possible, remove them early enough in the growing season so the tree has time to respond to the cut before going dormant. If you remove large branches later in the growing season or over winter, leave a stub instead of making a flush or concave cut. Leaving a stub can be a good idea when removing large branches any time of year because it lets the tree reroute the sap flow before you make the final cut that will heal over. You can remove stubs or use them to create deadwood features in the following growing season.

Stub cut and stub carved into a jin

Evaluating Your Results

The more familiar you become with how your trees respond to the work you do, the easier it will be to make your trees look the way you want them to look. After you've finished pruning a tree, evaluate your work by checking the following:

- Is the silhouette attractive?
- Is the branch or foliar density even from top to bottom when viewed from all sides?
- Are branch pads clearly defined?

You can ask additional questions after the tree puts on a new flush of growth in response to your work:

- Did new buds appear where you expected them?
- Are weak areas gaining vigor or getting weaker?
- Is the silhouette filling in as expected?

If weak areas continue to get weaker, thin branches that may be blocking their light. And if the silhouette isn't filling in as expected, make adjustments as necessary. For example, if a tree looks awkward or uneven after pruning, let weak areas grow stronger before pruning next time and/or prune strong areas more frequently. Repeating this process over time can create full, beautiful silhouettes.

MIDSEASON PRUNING (JAPANESE MAPLE)

Before we move on to wiring basics, here's a case study that shows how to put some of these ideas into practice. The tree is a Japanese maple that's been in training for several decades. It has an established shape but lacks branch density. The current development goal for the tree is to increase the branch density and refine the silhouette.

Early spring is a good time to check and see if deciduous trees need pruning, as they can grow quickly when they're healthy. Although the new leaves of this maple have yet to harden off, it's already clear that the leaves near the top of the tree are larger than the leaves on the lower branches. If the tree isn't pruned soon, the larger leaves will shade out the branches below and cause them to become weak.

Thinning dense branches and reducing the silhouette on the upper half of the tree will slow down growth and ensure the lower branches receive more light.

When selecting branches to prune, look for branches that are already too strong (opposite, upper right). After being pruned the previous year, this branch was strong enough to divide into four smaller branches. Instead of removing two of the new shoots and keeping the lower section of the branch (which is already too strong), removing the whole branch will allow a new one to develop with better structure (see page 82 for an example of a maple branch with better structure).

A good test to determine how much thinning is appropriate is to see if the trunk is visible through the foliage. If the leaves block your view of the trunk, sunlight is unlikely to reach the tree's interior, and any branches there will weaken or die back.

After I thinned the branches on the upper half of the tree and reduced the silhouette, it was easy for light to reach the tree's interior.

If a tree puts on an additional flush of growth, it can be thinned again in midsummer. Any growth that appears later than that can be removed in fall when the leaves turn color and begin to fall off.

Japanese maple

Strong branch with poor structure

Adam and Jonas pruning

After thinning

WIRING BASICS

Like pruning, wiring is a core technique used to make beautiful bonsai. If you're new to wiring, you may find that doing it well requires a lot of practice. The best way to gain this practice is to find trees that need styling. If you run out of bonsai to work with, you can cut branches from garden trees, clamp them to your workspace, and wire them. And if buying more wire is cost prohibitive, reuse aluminum wire by removing it from practice branches, straightening it, and reapplying it to new branches.

For your wiring to have meaning, it must be effective. If your wiring fails to hold a branch in the desired position, remove it and try again by rerouting it or by using a different gauge wire. If possible, get feedback about your wiring from a professional or from an experienced hobbyist whose work you admire. The sooner you can identify opportunities for improving your wiring, the sooner you'll form effective habits.

Well-anchored wire

Selecting the Right Kind of Wire

The most common materials for bonsai wire are anodized aluminum, annealed copper, and stainless or galvanized steel. Each has different characteristics that make it better suited to some tasks than others.

Aluminum wire is softer than steel or copper, which makes it a good choice for wiring soft or brittle branches. As such, it's a popular choice for wiring deciduous trees. It comes in brown, black, or its natural silvery color. Because it's softer than copper, a larger gauge wire is required to apply the same holding power as a smaller gauge copper wire.

Copper wire is stronger than aluminum but softer than steel. It's a popular choice for wiring conifers because it has more holding power than aluminum and facilitates bigger bends. Over time, copper turns black, which helps it blend in with the relatively dark-colored bark of conifers.

Steel wire can be used to secure heavy trees, or trees set at severe angles, into their containers. Steel wire is generally not used to wire branches, but it can be a good choice for guy wires when additional strength is required.

Selecting the Right Size Wire

There's no set rule for how to select an appropriate gauge wire to bend a given branch. Finding the right size will depend on the flexibility of the branch and the severity of the bend. For supple branches, the diameter of the wire may be as small as 25 percent of the diameter of the branch, particularly if you're using copper wire on a small branch. If you're using aluminum wire on a heavy branch, you may need to use wire with a diameter closer to 50 percent (or more) of the diameter of the branch.

If you select a gauge that's too light, the branch won't stay put when you move it. And by selecting wire that's too heavy for a branch, it's easier to damage or break the branch when it's time to set the bends. Whenever you discover that a wire is too small (or too large) for a given branch, remove it and try the next gauge up (or down) so you can safely move the branch into place and keep it there until the desired bends set.

Effective Body and Hand Positions for Wiring

When wiring, work at chest height as much as possible, with your upper arms relaxed and your forearms slightly forward. This will give you good leverage as you apply the wire and help you remain comfortable while you work.

Apply the wire with your dominant hand. Use your nondominant hand with your palm facing upward to secure the wire against the branch. If you see gaps in your wiring or the wire you apply isn't snug, try pinching the wire more tightly against the branch with your nondominant hand as you work.

After applying a coil of wire, move your nondominant hand to the most recently applied coil and continue wiring to the end of the branch. If you find wire challenging to apply, particularly with heavier gauge wires, try holding the wire farther out to give you more leverage. Using pliers (and/or tweezers on smaller trees) can also provide additional leverage and can help you apply wire in hard-to-reach areas.

In general, coil wire around branches at an angle of approximately 60 degrees. If you want to create tighter bends, apply wire at angles closer to 45 degrees. Ensure that the wire is snug, but not tight, against the branch; wire that's too tight can cut into the bark quickly.

If you find it challenging to create consistent coils, try twisting the wire slightly as you work to give the wire a slight curve as it makes contact with the branch.

Securing the wire

Sequencing the Work

In general, begin work with the lower branches and progress upward to the branches that form the apex. Because wire is typically used to bend branches down, wiring the lowest branch first creates space for you to lower the next branch above it, and so on, as you work.

Likewise, apply larger wires before applying smaller wires. This way the larger wires can provide anchors for the smaller wires.

Whenever possible, try to connect two branches with a single wire. Wiring branches in pairs simplifies the wiring process by reducing the number of wires you need to apply. You can check to see if you've paired branches correctly by making sure the wire on each branch spirals in opposite directions.

Applying wire at chest height

Anchoring Your Wire

The best way to ensure your wiring is effective is to learn how to anchor your wire.

Anchoring is accomplished when a wire relies on an immobile part of the tree to stabilize the portion of a branch or trunk that you'd like to bend. You can use two tests to determine whether or not a wire is well-anchored. If you adjust the position of one branch and another branch moves, the wire is not adequately anchored (these are known as "teeter-totter branches"). You can also tell if a branch isn't anchored properly if you use the right size wire but it doesn't stay put when it's bent.

You can anchor your wire by connecting or wrapping a portion of it around an immobile feature of the tree such as the trunk, the base of a branch, or a deadwood feature. You can also anchor wire to other well-anchored branches. Here's a simple example that demonstrates how this works.

Wire A is the first wire applied to the branch. Wire B is added next, followed by wires C and D. The zone where wires A and B run next to each other is the zone where wire B is anchored by wire A. Likewise, wire B anchors wires C and D where they overlap.

This pattern of using the end of one wire to provide an anchor for the next wire is the most common way to anchor your wire. If you can learn this pattern, you can wire the majority of branches on your trees.

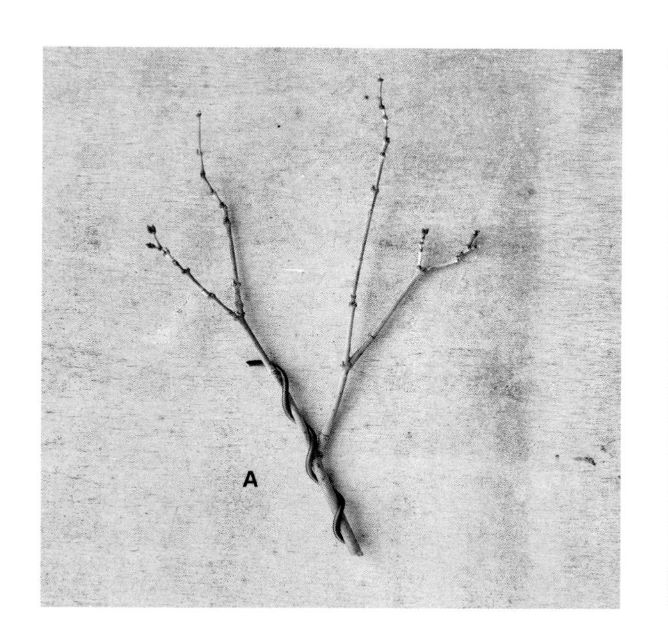
A

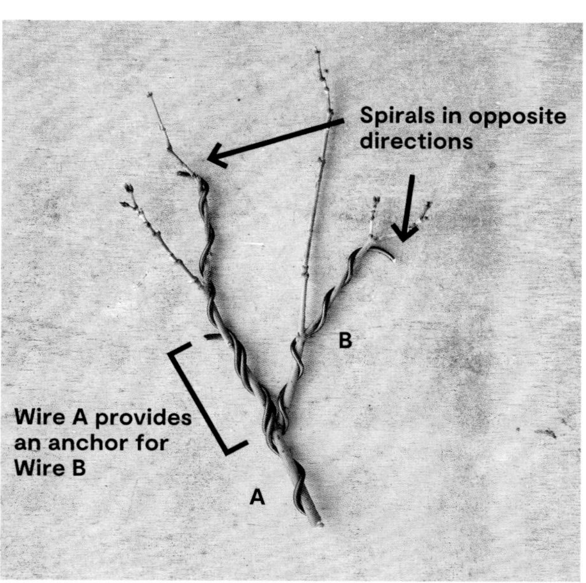
Spirals in opposite directions
B
Wire A provides an anchor for Wire B
A

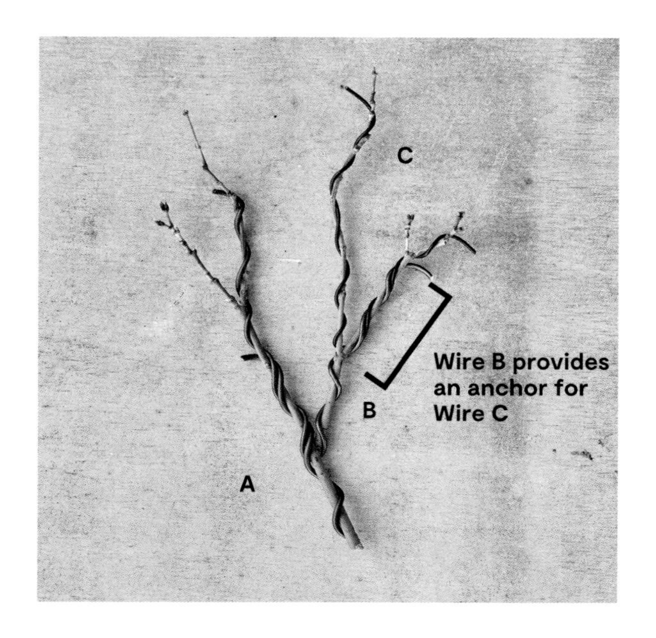
C
Wire B provides an anchor for Wire C
B
A

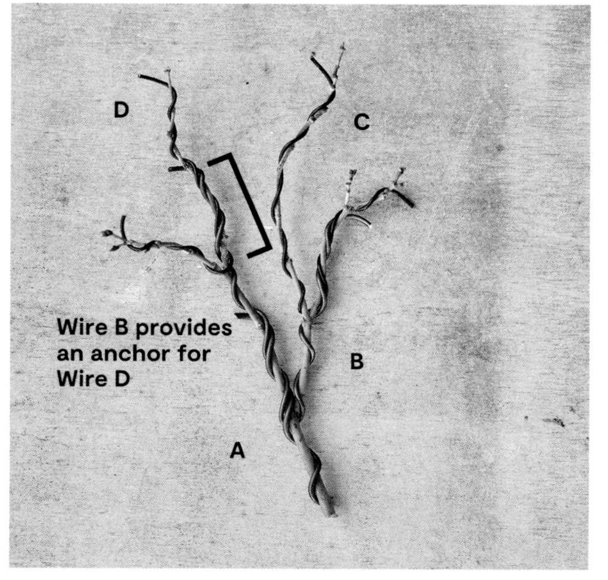
D
C
Wire B provides an anchor for Wire D
B
A

Setting Branches

Setting a branch is the act of shaping and/or repositioning a wired branch. When you completely wire a tree, setting the branches is often the final step before you return the tree to the garden.

When you're setting branches, work with the tree at eye level to make sure it looks good from the designated front. Start by bending the base of the branch and work your way out to the branch tips, making one bend at a time. Set the lower branches first (just as you did when you applied the wire) and work upward toward the apex. Some people set branches as they work, while others apply all of the wires first and adjust them later. Either approach, or a combination of the two, is fine.

Setting branches at eye-level

A good way to assess your work is by asking a few questions about the consistency of your wiring:

- Are the angles between the primary branches and the trunk consistent (mostly upward or mostly downward), or do they appear arbitrary?
- Is the branch movement consistent (do all branches, for example, have subtle bends), or do different branches look like they belong to different trees?
- Do the spaces between branches appear natural, or are there awkward gaps? (Adjusting branch angles can help you close unwanted gaps.)

You can ask similar questions about the pad design, silhouette, and flow. (Does the shape, size, and arrangement of branches complement the trunk? Is the silhouette in scale with the trunk? Does the tree clearly flow in the direction you want it to?)

Another option for evaluating your work is to take a picture of the tree and study the work in two dimensions. To get the best view of the tree, take a few steps back from it and hold the camera at a height of roughly one-third of the way between the top of the pot and the top of the tree.

When setting branches, it's okay to move them into different positions to test which options are best, but doing this too much can weaken branches and lead to dieback.

Avoiding Ineffective Wiring Patterns

The best way to avoid bad wiring habits is to learn to recognize flawed patterns so you can remove these wires and reapply them correctly. It may require several attempts to get it right, but learning to apply wire effectively will help you avoid breaking branches and give you more control when it's time to bend them.

The Image Below Shows Examples of Poor Wiring Techniques

1. Uneven spirals
2. Wire isn't anchored
3. Gap at branch shoulder
4. Gap at base of wire
5. Crossing wires
6. End of wire loops back too far
7. Wire too short
8. Wire gauge too heavy

How to Determine When to Remove the Wire

You'll know that it's time to remove bonsai wire from a tree when a branch can hold its new position or when the wire cuts into the bark. It's difficult to determine whether branches have set until you start removing wire, but if it's been more than a year or two, it may be worth a test to see if the branches stay put. After removing a few wires, wait a week or two to see if the branches hold their shape. If they do, you can remove more wires, and if they don't, you can rewire any branches that need it.

It's far more common to remove wire because it's causing swelling that can lead to unsightly scars. This can happen in a matter of weeks on young trees that are growing quickly or over a period of years on slow-growing conifers. Swelling rarely affects branches evenly, so check wires carefully to avoid unwanted scarring. This could mean weekly checks on deciduous trees in spring or monthly checks on conifers.

Although some swelling is acceptable on conifers, it's not acceptable when the swelling is distracting or when it affects the integrity of the branch. Wire scars on the trunk are never acceptable. For deciduous species, the goal is to leave no wire marks whatsoever because these trees have no leaves to hide scars in winter. If only some wires are causing swelling, it's okay to remove only those wires and leave the others in place. When the majority of wires are starting to cut in, however, it's best to remove all of the wires on the tree and rewire as necessary.

In general, remove wire from the tree in the opposite order in which you applied it. Remove smaller wires first (by hand) and work up to the largest wires last. You can remove large wires by cutting one spiral at a time with a wire cutter.

PUTTING IT ALL TOGETHER—PRUNING AND WIRING A JAPANESE MAPLE BRANCH

Maple branch before pruning—top

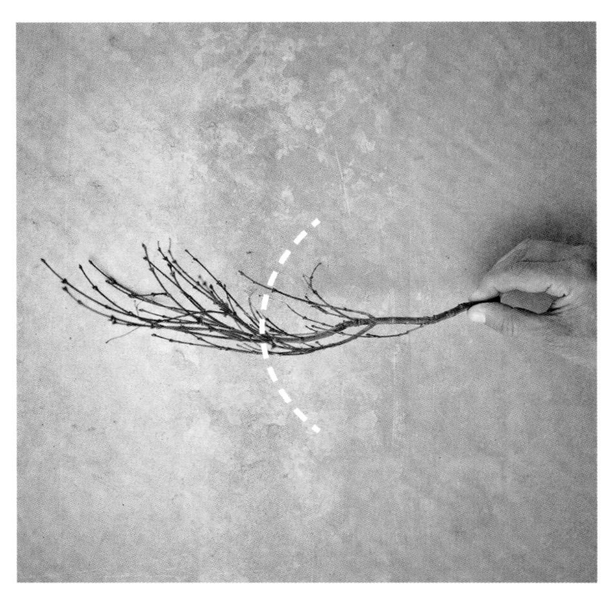

Before pruning—side

After pruning—top

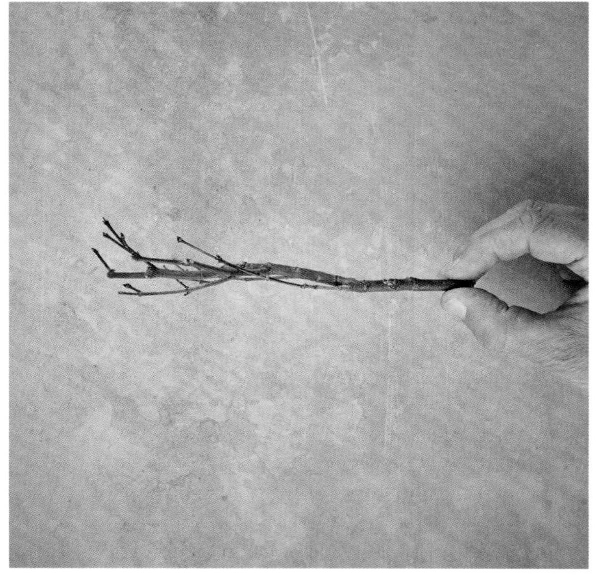

After pruning—side

Now that you have an overview of basic pruning and wiring, let's look at an actual example. The branch shown in the top photos at left is from a Japanese maple that has grown out of shape and needs significant cutback. The goal is to improve the branch structure by pruning and then wiring the smaller branches into a basic fan shape.

IMPROVING THE BRANCH STRUCTURE

- Identify a central leader for the branch and prune it to the desired length.
- Establish the shape of the branch by reducing side branches.
- Remove side branches when more than one emerges from the same location to create an alternating branch pattern (one branch on the left followed by one on the right, and so on).
- Apply wire to the branches. The heaviest wire is applied first to the central leader. This wire provides an anchor for the next largest wire, which in turn provides an anchor for the smallest wires.
- Set the wired branches to create a basic fan shape.

To get an idea of what different species look like before and after wiring, see the case studies on pages 186 (olive), 188 (cryptomeria), and 192 (white pine).

WIRING CHECKPOINTS

- Branches are wired in pairs.
- Spirals are consistent.
- Larger wires provide anchors for smaller wires.
- Wire gauge is appropriate for the branch size.
- Side branches emerge from the main branch at acute angles.
- Branch tips point away from the main branch line.
- There are no crossing wires.
- There are no more than three wires next to each other on any given branch (too many wires in a given area can indicate poor routing decisions).
- Wires have a consistent finish (the ends of the wire aren't too long or too short; aim for the final curve of wire to support each branch with a gentle curve as pictured below).

After wiring

CHAPTER 5

Pinching, Leaf Cutting, Defoliating, and Decandling

Some species trained as bonsai require more than just pruning and wiring to become beautiful. If a tree you're working with lacks branch density or fails to produce fine, twiggy growth, techniques such as pinching, leaf cutting, defoliating, and decandling can help. This chapter provides starting points for applying these techniques so you can create beautiful branches that give your bonsai character as they mature. The techniques are presented in the order in which you can expect to perform them throughout the growing season: pinching and leaf cutting in spring, defoliating and decandling mid-season, and needle thinning in fall.

PINCHING

Pinching refers to the removal of new growth with scissors, tweezers, or your fingers. It's a technique for reducing vigor and stimulating budding for trees in middle or later stages of development, when the goal is to improve branch structure and density. (Young trees generally aren't pinched, as the goal at this stage is to speed up growth, not slow it down.) The approach to pinching differs depending on the species, so we'll look at pinching deciduous trees, pines, and coast redwoods separately.

Pinching Deciduous Species

Pinching is a common technique for Japanese maple, trident maple, zelkova, hornbeam, and Japanese beech, as these species naturally produce long internodes when they grow freely in spring. The right time to pinch terminal buds on strong branches is when new leaves begin to emerge. Because the buds on deciduous trees typically open over a period of days or sometimes weeks, these trees can require daily attention throughout early spring.

The photos at right show the steps for pinching new growth on a Japanese maple.

In general, pinching slows branch growth. If, after pinching, new growth continues to be strong, you can continue pinching throughout the growing season. If you're not sure whether or not a species is a good candidate for pinching, try it out in strong areas on relatively young and healthy trees before experimenting more broadly.

New leaves to be pinched

Grabbing the leaves to be pinched

After removing the new leaves

Pinching Pines

When pines start to grow in spring, new buds or shoots known as "candles" begin to elongate. If your goal is to promote balanced growth across the tree, you can break or pinch the longest candles before they mature.

For pines such as Japanese black pine or red pine, you can pinch candles after they begin to elongate but before new needles emerge. Reduce the longest shoots by two-thirds. Reduce midsized shoots by half. There is no need to pinch small shoots because they are already a good size for the tree.

For high-mountain pines such as ponderosa pine, white pine, and mugo pine, you can pinch new growth on strong branches later in the growing season, just before it hardens off in spring. This approach also applies to cedar and spruce bonsai.

Elongating shoots or "candles"

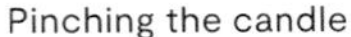

Pinching the candle

After pinching

Pinching Coast Redwood

Pinching redwoods is a good way to build foliar density and shorten internodes. The basic approach is to remove a portion of new leaves with fingers or scissors before they harden off. If the foliage is tender when you pinch, there is a good chance the leaf will produce buds near the cut site. If you pinch a redwood leaf after it hardens off, the remaining portion of the leaf is likely to die back.

The main benefit of pinching a redwood is that the technique offers great control over internode length, an important tool when creating a composition that suggests a much larger tree.

Mature leaf, tender leaf after pinching, new buds on a leaf that was pinched

LEAF THINNING AND LEAF CUTTING

Japanese maple branch

After removing every other leaf

After leaf cutting

Before and after—different approaches to leaf cutting

Over time, as the branch density on your trees increases, you may find that it's difficult to maintain the health of small branches in the tree's interior. If this area doesn't receive adequate light, the branches there can become weak or die. To keep interior branches healthy, you can remove leaves, or portions of leaves, from the outer canopy.

Leaf thinning is a common way to reduce vigor and let light into the interior of dense Japanese maples. In strong areas or areas of the tree that have grown too dense, remove every other leaf.

If the area you're working on is still too dense, you can further reduce the foliage by cutting the leaves.

There are multiple ways to cut the leaves of Japanese maples (see opposite).

On species such as stewartia, zelkova, Chinese quince, and beech (see below), strong leaves can grow much larger than weaker leaves. You can make leaf size more consistent and let light into the tree's interior by cutting the larger leaves. Folding a leaf in half before cutting can preserve the leaf shape.

If your tree grows well after thinning or cutting the leaves, you can repeat the process as needed on new foliage.

Small leaf, immature leaf after leaf cutting, mature leaf after leaf cutting

Leaf cutting European beech

DEFOLIATING

Korean hornbeam

Defoliation is a technique for removing some or all of a tree's leaves to reduce or balance vigor on selected deciduous and broadleaf evergreen species. Full defoliation involves removing all of a tree's leaves. It's a stressful operation aimed at slowing a tree's growth. As such, it's applied only to fast-growing species, such as trident maple and silverberry, that respond well to the technique. Partial defoliation is a technique for removing a portion of the tree's leaves. Partial defoliation reduces vigor in strong areas (such as the outer canopy) while letting weak areas gain strength. Both techniques provide additional light to branches growing in the tree's interior.

Here's an example of how the technique can work. The tree shown at left is a mature Korean hornbeam. The first flush of spring growth has nearly hardened off, but the branching is so dense that some interior leaves are already turning yellow. If these leaves don't receive additional light, it's likely that some small interior branches will die.

One reason the tree is so dense is that the new spring shoots are 1 to 1½ inches long. To see if shortening the spring growth will let adequate light reach the tree's interior, I cut all of the new shoots back to two or three leaves (about ½ inch).

After pruning—front

After pruning—top

As you can see, the foliage is still dense after shortening the new branches. When viewed from the top, there are still too many leaves to allow light into the tree's interior. This is a perfect opportunity for partial defoliation. Here's what the tree looks like after removing the majority of the leaves growing in the outer canopy.

It's now easy for light to reach the interior branches. Over the coming weeks, new leaves will develop in the outer canopy while giving the interior branches enough time to gain vigor and maintain health until the leaves turn color in fall.

Partial defoliation can also work branch by branch. If most of the tree is strong and one or two branches are weak, you can partially or fully (depending on the species) defoliate the strong areas and leave the weak branches alone to give them a chance to gain strength.

After partially defoliating—front

Species that respond well to full defoliation include trident maple and silverberry (up to two or three times a year) and young or very healthy Japanese maples (once per year). Species that respond well to partial defoliation include Chinese quince, cork oak, hornbeam, Japanese maple, ume, and zelkova. Species such as beech and stewartia are good candidates for leaf pruning instead of partially defoliating, as defoliating these species can lead to branch dieback.

After partially defoliating—top

DECANDLING

1. Shoot to be decandled

2. After decandling

3. Shoot to be cut back

4. After cutback

Decandling is a technique for balancing vigor and building dense branches on selected pine species (see page 91 for a list of compatible species). When pines start growing in spring, the buds, which resemble candles, elongate and produce new needles. If you remove this growth in the middle of the growing season, a healthy pine will produce a second flush of growth that is more compact than the spring flush.

Performing the technique is simple. In the middle of the growing season, remove the spring growth, being sure to leave a small stub of new tissue where you make the cut, as this is where the new buds will develop (photos 1, 2 at left).

You can do this to every new shoot on a healthy pine. Here's what this looks like on a relatively mature specimen.

A variation on the technique involves cutting back not only the current year's shoots but also some of the previous year's growth (photos 3, 4 at left). Instead of leaving a small stub, be sure to leave a few pairs of healthy needles, as the goal is to stimulate needle buds (buds that develop at the base of a pair of needles). This approach can help you reduce the internode length of the *previous year's* new growth.

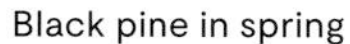
Black pine in spring

After decandling

You can control the length of the needles that develop over summer by modifying the date on which you decandle. Here are some tips to help you produce needles that are the right size for your tree:

- The standard time to decandle pines is around the summer solstice. If you live in cool or northern cities such as Seattle, you can start decandling about a month before solstice. If you live in warm or southern cities such as Atlanta, start decandling a week or two after solstice.
- To produce the exact size needles you're looking for, adjust the timing based on your results from the previous year's decandling. For example, if needles were too long, try decandling a week or two later. If they were too short, try decandling a week or two earlier.
- Decandle your larger pines first (so the needles have more time to elongate during the growing season) and your smaller pines two or three weeks later (so they have less time to elongate).

You can keep pines in full sun after decandling, but you may find they need watering less frequently until they produce new foliage. Fertilize pines well before decandling. You can pause fertilizing at decandling time (to prevent summer shoots from growing too strong) and start fertilizing again in fall. Finally, you don't have to decandle every branch on a tree. If only a few branches are strong, you can decandle those and leave the other branches alone.

If you cut some of the previous year's growth when you decandle (photos 3 and 4 on page 88), a few additional tips apply:

- When working with young, healthy trees, it's okay to apply this technique to every strong branch on the tree. When working with mature trees, the technique works best on strong shoots that receive ample sunlight.
- This technique doesn't work as well for trees that have recently been repotted or have weak root systems.
- If the tree produces buds but no new shoots, this is okay because these buds may open the following spring. If any branches fail to produce new buds, it's a sign they weren't strong enough to respond well to the technique.

Keeping notes about your successes and misses can offer valuable feedback for making adjustments the following year. After a few years of practice, you'll be able to determine the best approach for your pines every time.

Lauren decandling a black pine

Which Species Can I Decandle?

Decandling works well for vigorous pine species including Japanese black pine, Japanese red pine, pitch pine, loblolly pine, slash pine, Virginia pine, longleaf pine, and shortleaf pine. Less vigorous pine species including cork bark Japanese black pine, dwarf Japanese black pine cultivars, Monterey pine, Scots pine, Aleppo pine, and Italian stone pine can benefit from partial decandling, decandling every other year, or limiting decandling on young or healthy specimens. Species such as Austrian black pine, pinyon pine, and mugo pine don't reliably produce summer growth after decandling, but pinching or removing strong shoots may be useful to stimulate back buds. Avoid decandling Japanese white pine, ponderosa pine, limber pine, bristlecone pine, foxtail pine, western white pine, eastern white pine, and shore pine/ lodgepole pine. For these trees, the standard approach to branch development is to pinch new shoots in spring and prune in fall or winter.

Summer growth a few weeks after decandling a black pine

NEEDLE THINNING

Needle thinning refers to the removal of foliage on pines for the purpose of balancing vigor, letting light into the tree's interior, and stimulating budding. Like decandling, it's a flexible technique that you can modify to suit your needs. Although the basics apply to a variety of pines, the approach noted here applies primarily to species that are decandled, such as Japanese black pine and Japanese red pine.

The best time to pull needles is in fall or winter. When necessary, you can also remove needles at decandling time. The most basic approach is to remove needles from strong shoots so that all shoots on the tree have approximately the same number of needles. Remove needles with your fingers or tweezers, being careful not to tear the bark where each pair of needles is attached.

Black pine that needs thinning

For full, mature pines, keep as few as three to five pairs of needles per shoot. For trees in earlier stages of development, keep up to ten to fifteen pairs of needles per branch. When you're getting started, remove a few needles on a test branch or two and see how it looks before thinning the rest of the tree. As you work, remove the oldest needles first and work your way toward the end of the shoot, keeping the needles that grow closest to the terminal bud. Remove yellow or brown needles.

Like defoliation, removing pine needles from a branch weakens it. If a branch or area of a branch is too strong, consider pulling more needles. If it's weak, avoid pulling any needles and ensure that the branch gets good sunlight by thinning needles from branches above it that may be blocking its light.

Needle thinning is a great way to balance vigor on pines. You can keep more needles on any areas of the tree you want to grow stronger and fewer in areas you want to slow down. For this reason, it's common to pluck more needles from the upper branches and keep more needles on the lower branches.

When you finish a first pass at needle thinning, evaluate your work by taking a few steps back from the tree and looking at it from all sides to make sure the work is consistent with your goals for thinning. Two to three weeks after pulling needles, check the foliage for any signs of damage. If you see bent, broken, or discolored needles, work slowly and carefully next time to avoid damaging healthy foliage.

As with decandling, take note of how your trees respond to needle thinning and adjust as necessary the following year.

Strong, medium-strong, and less-strong shoots

After removing needles from the strong and medium-sized shoots; no needles were removed from the weakest shoot

Cork bark pine after decandling and thinning needles

ERIC SCHRADER

Whenever I see something that needs doing in the garden and my excuses for not doing it sound flimsy, I think about how Eric Schrader would approach the work.

As a hobbyist, Eric maintained about 100 bonsai. Now he and his wife, Dory, run a much larger operation through their business, Bonsaify. The number of trees Eric can work on in a day is staggering. His secret? Among other things, great habits.

Developing beautiful bonsai requires horticultural knowledge, artistic sensibility, and technical skills. It also requires good habits. Beautifully styled trees don't fare well when the watering is irregular. And even the most skilled pruning can be counterproductive if it doesn't happen at the right time of year. As Eric says to his students, "In bonsai you're always working on the tree, but what are you doing to work on yourself so you can make better trees?"

On a daily basis, developing good habits may mean spending more time with your trees to assess water needs and identify changes in health or signs of infestation. On a weekly basis, it could mean working to keep your bonsai free of weeds, rotating trees so they receive equal sunshine on all sides, and applying fertilizer as needed throughout the growing season. On a seasonal basis, you may need to ensure that you're pruning when new growth gets out of hand, repotting trees that require it, and taking the time to restyle trees that are ready for a change.

How can you find time to make these incremental improvements and improve your knowledge? Here are some tips:

- Create a dedicated space for working on bonsai to reduce the time you spend setting up, looking for tools and supplies, and putting everything away afterward. If you can spare only an hour a day to work on your trees, try to spend as much of that time as possible maintaining your trees rather than maintaining your workspace.
- Spend time with people who create or maintain beautiful trees so you can learn from one another, share successes, and offer (or receive) help when needed.
- Commit to showing your best trees in a local or regional exhibit. Working toward a goal will help you learn everything you can do to make your trees look their best at a specific time of year.

SEASONAL TASKS

A number of bonsai tasks require your attention throughout the year. When you see that the following conditions need attention, it's time to do the work.

Remove shoots that emerge directly from the soil or from the base of the trunk. Unless you are creating a multiple-trunk or clump-style composition, these shoots will gain vigor over time and can cause the upper portion of the tree to become weak.

Remove moss growing on the trunk or surface roots because it can erode or discolor the bark. If the bark is flaky, take care to avoid damaging it when you remove the moss.

Remove spent flowers to prevent fruit from forming. For species such as azalea or wisteria, be sure to remove the entire flower (not just the petals) to prevent seeds or seed pods from maturing.

If you can't remove the flowers in time, remove the resulting fruit at the earliest opportunity.

Remove female cones on species such as pine when they are big enough to twist or break off with your fingers. Let pollen cones mature and dry out before brushing them away with your fingers or tweezers. For species that produce substantial cones that don't come off easily (such as cedar), remove cones with scissors or cutters.

Removing basal growth on coast redwood

Removing fruit from chojubai

Moss growing on the trunk

After removing the moss from the trunk

Azalea flower to be removed

After removing the flower

CHAPTER 6

Repotting

Repotting is one of the most important, but least understood, bonsai techniques—in part because we don't get an opportunity to see the roots up close on a regular basis. Unlike pruning or wiring, where our overall goal is to make the tree look beautiful, the overall goal of repotting is to maintain tree health. This is why repotting is among my favorite bonsai activities—it's an opportunity to work with the hidden parts of the tree and facilitate watering, fertilizing, and overall tree health.

SIGNS THAT BONSAI NEED REPOTTING

Water pooling in the pot

Repotting is a routine part of the bonsai lifestyle. We repot bonsai to maintain tree health and to provide trees with containers that support our goals for development or display.

An obvious sign that a tree needs repotting is when the roots push the root ball out of the pot. You can also repot for aesthetic reasons such as changing the front, the growing angle, or the container the tree is growing in. If a tree is new to your collection, you can repot to learn more about the root health or to search for and remove pockets of old soil within the root ball.

You can perform several tests to determine whether a tree needs repotting:

- Press down on the surface of the soil to see if it has any give to it. If the soil is firm to the touch, it may be time to repot.
- Insert a chopstick or similar object into the soil and see how far down it reaches. If it is difficult for the chopstick to penetrate the soil, it may be time to repot.
- Water the tree and see how long it takes to drain. If the water pools before draining, it may be time to repot.

The third test is the most important. If you find that it's difficult to water a tree or you need to fill the pot several times before enough water percolates through the soil to wet the entire root ball, it's time to repot.

In all of these cases, soil particles breaking down and/or lots of root growth create the need for repotting. By reducing the roots and replacing at least a portion of the old soil when you repot, you can provide the tree with more favorable conditions for root growth.

Avoid repotting bonsai on a schedule unless you're familiar with how the tree responds to different repotting frequencies. In most cases it's best to repot when the above signs indicate that it's time. For example, young trees may need repotting every one to three years, on average, depending on how fast they're growing or what kind of soil they're growing in. For well-established bonsai, repotting every three to five years, or longer, may be appropriate.

Roots pushing the tree out of the pot

SEASONAL TIMING OF REPOTTING

Repotting season for bonsai is primarily determined by the weather. For trees that go somewhat or fully dormant in winter, the best time of year to repot is when they start growing in spring. This applies to most deciduous species, conifers, and non-tropical broadleaf evergreens that are commonly trained as bonsai. For tropical species such as jade, ficus, or buttonwood, you can repot in summer when the trees are actively growing.

Because some species start growing earlier than others, repotting season can be as brief as a few weeks or as long as several months.

By getting the timing right, you can perform the most root work while causing the least amount of stress to the tree. The main reasons to repot at other times of year are when a container breaks or a tree is unhealthy due to extremely poor soil conditions. In these cases, do less invasive root work to avoid overstressing trees.

Scoria, pumice, akadama

Perlite, fir bark, mulch, kiryu, kanuma

BONSAI SOIL INGREDIENTS

Akadama is a claylike volcanic particle that's mined in Japan. It's valued as a soil component around the world for its ability to facilitate healthy root growth. Over time, akadama particles break down into smaller particles. Some of this degradation is beneficial because it can promote fine root growth. Too much degradation, however, can make watering difficult and lead to root problems, a key reason why we repot bonsai periodically.

Pumice is a lightweight volcanic rock characterized by its pitted and cavity-filled form. Unlike akadama, pumice can provide long-term structure to the soil because it doesn't break down over time. It has a lower water-holding capacity than akadama, scoria, perlite, or organic compounds such as bark or mulch. If sifting doesn't successfully remove excessive dust, try rinsing it with water and letting it dry before use.

Scoria (also known as cinder or lava rock) is a volcanic rock, like pumice, characterized by its pitted and cavity-filled form. Scoria is darker than pumice (it's typically black, brown, or red) and denser (some pumice can float, while lava sinks). The characteristics of scoria are determined by the material's mineral composition and stage of decomposition. In general, scoria is harder than pumice and can provide structure to the soil over long periods of time. Rinsing and letting it dry before use is highly recommended to wash away excessive dust, when present.

Perlite is a lightweight volcanic rock made by heating obsidian. It's softer than pumice and can hold up to three or four times its weight in water. Because of its light weight (particles can easily wash out of the container) and bright white color (which some find unappealing for bonsai), perlite is more popular for seedlings, cuttings, and pre-bonsai than it is for mature specimens.

Organic soil ingredients including mulch, bark, coir, and peat are popular in the nursery industry and used widely for developing pre-bonsai. It's far less common to use organic material in mixes for mature specimens, as these ingredients can break down relatively quickly, compared with volcanic ingredients, and retain excessive moisture.

Kiryu or Kiryuzuna is a yellowish volcanic rock mined in Japan that's slightly acidic like kanuma but hard like pumice. It's a common ingredient in Japanese premixes.

Kanuma is a volcanic rock that forms in deposits immediately below the akadama layer. It's similar to pumice with a few key differences: it's softer, yellower, and slightly acidic. Kanuma's acidity makes it a popular choice for growing acid-loving species such as azaleas and camellias.

CREATING AN EFFECTIVE SOIL MIX

Poorly sifted soil and well-sifted soil

When selecting ingredients for a soil mix, consider cost, availability, and how well it works for your trees. That's it!

A simple and popular mix for conifers is one part pumice, one part scoria, and one part akadama. For deciduous trees, it's one part pumice, one part scoria, and two parts akadama. If scoria isn't available in your area, you can replace it with more pumice, as both ingredients play a similar role in the soil.

These mixes have been used across North America with good results, but don't hesitate to make changes if they aren't working for you. In general, use smaller particles (up to ¼ inch in diameter) for small and medium-sized trees and larger particles (up to ½ inch in diameter) for larger trees. If your trees dry out too quickly, try using smaller particles or a higher percentage of akadama. If your mix stays too wet, try using slightly larger particles or less akadama the next time you repot.

You can also change the composition of your mix based on the needs of the species you grow. For example, coast redwood, hinoki, and cryptomeria are conifers that prefer soil mixes that retain more water. As a result, you'll find it easier to keep these species healthy in mixes with higher percentages of akadama (up to 50 percent, 70 percent, or more, depending on your climate).

You may also find that the same tree can benefit from different mixes over time. Young trees and rooted cuttings are well suited to perlite (70 percent) and coir or peat moss (30 percent) mixes, but once the trunk development is complete and you begin to focus on refining the branches, you may find it easier to achieve your development goals with akadama-based mixes.

Whatever mix you go with, use particles that are roughly the same size when you combine ingredients, and sift out any dust that may have accumulated. Poorly sifted or dusty mixes can retain excessive moisture, make it difficult to keep trees healthy, and reduce the time before your trees need repotting again.

It can be helpful to remember that no single ingredient is required for any given species. Bonsai will grow without akadama, without organics, and even without volcanic ingredients altogether. I like to experiment with different ingredients and use whichever ones make it easiest to keep my trees healthy and facilitate my development goals. For my mature bonsai, this usually means mixes with akadama, lava, and pumice. For my younger trees, it can mean just about anything that produces good results.

When you make it to the end of repotting season, you may wonder if it's okay to reuse old bonsai soil. In general, the answer is yes.

TIPS FOR REUSING OLD BONSAI SOIL

- Dry used soil in the sun until it's completely dry.
- Sift dried soil to remove dust and other small particles.
- Add ingredients that may have broken down over time such as akadama or organic components to get the soil back to your target percentages.
- Use caution when reusing soil harboring insects such as root aphids or fungal pathogens, as this can increase the odds that you'll spread the problem to other trees.

SELECTING CONTAINERS FOR DIFFERENT STAGES OF DEVELOPMENT

Plastic, wood, and ceramic containers

When it comes to selecting containers for your trees, there are a lot of choices, especially for bonsai in early stages of development.

A common mistake is to use containers that are too small. Although it can be satisfying to find the "right pot" for a given tree, **using slightly larger containers until the tree has the silhouette and branch density you're aiming for can accelerate development and make it easier to keep your trees healthy.**

There are also good reasons why you might want to use different kinds of containers for trees at different stages of development. Here are some suggestions for how this can work.

COMMON CONTAINER-TREE PAIRINGS

- Plastic containers used by the nursery industry are good for young trees or trees in early stages of development, as the shape and size of the pot matters less while the tree is still taking shape.
- Containers such as pond baskets, colanders, grow bags, and Anderson nursery flats are great for facilitating rapid trunk development and fine root growth for species like pines that like relatively dry root environments. These perforated containers provide additional airflow to the root ball, which prevents roots from spiraling and allows for more frequent watering without keeping the soil too wet.
- Wood boxes are a natural choice for recently collected trees (they can be custom built to fit perfectly) and species that prefer growing in moist environments (including maples and other deciduous species).
- Terra cotta pots are a good choice for trees that are starting to look like bonsai, as they are roughly similar in size and shape to bonsai containers.
- Ceramic bonsai containers are appropriate when trees reach their intended size and shape. This is usually the stage at which your focus shifts from trunk or primary branch development to fine branch development.

HOW TO REPOT

The remainder of this chapter focuses on how to repot, how to care for recently repotted trees, and how to evaluate your results. Before getting started, mix and sift any soil you'll be using so it will be ready when you need it

Repotting Steps

The following photos demonstrate the steps for repotting a Japanese black pine.

1. Remove the tree from the container.

   Remove moss, if any, growing on the surface of the soil.

   Cut any tie-down wires securing the tree to the pot.

   Remove soil from along the pot wall with a chopstick or sickle. Clear the soil from at least two short sides and one long side of the container. Working the sickle down to the bottom of the pot will make it easier to remove the tree without damaging the root ball.

   Tilt the tree out of the pot.

Japanese black pine

Removing moss

Cutting tie-down wires

Tie-down wires removed

Removing soil along the sides of the pot

Tilting the tree out of the pot

2. Prepare the container.

Scrub the container gently with a wet rag or soft brush to remove any dirt. Be careful not to remove patina that may have built up over time.

Secure mesh to the pot with wire in order to cover the drainage holes.

Insert pre-bent tie-down wires (see "Securing Bonsai in the Pot," page 120).

If you don't know which container you'll be using, line up different options ahead of time, so you can test each after completing the root work.

Scrubbing the pot

Inserting drainage mesh

Inserting tie-down wires

3. Perform the root work.

   Determine the level of the surface roots by removing soil from the top of the root ball until you can see the zone where the trunk transitions into lateral roots. (Skip this step if you can see the surface roots clearly.)

   Place the tree on its side and use a root rake or chopstick to comb out the mat of fine roots on the bottom of the root ball. As you work, make sure the surface is flat rather than concave to avoid creating air pockets when you set the tree in the pot.

   Trim roots on the bottom of the root ball. Make sure a small portion of the roots protrudes from the root ball about ⅛ inch to ¼ inch, as this helps introduce the roots into new soil.

Combing out roots on the bottom of the root ball

Trimming the bottom of the root ball

Use root cutters to remove large roots.

When work on the bottom of the root ball is complete, the tree should sit at the angle it will be growing in the container.

Removing a larger root with root cutters

Root work on the bottom of the root ball is complete

Remove the top layer of soil with a chopstick or root hook. Be careful not to damage surface roots.

As you work, remove any dead or unattractive roots such as crossing roots, roots with unsightly bulges or scars, and roots that emerge from the trunk too far above or below the other roots. It's generally safe to remove up to one-third of flawed surface roots in a given repotting.

Trim long roots that extend from the sides of the root ball.

Aim to make the root ball roughly the same shape as the pot it is going into but smaller so there's space for new soil.

Combing out roots on the sides of the root ball with a root hook

Removing old soil from the top of the root ball

Trimming the sides of the root ball

Perforate the root ball with a chopstick or metal pick to loosen compacted areas of the root ball and make space for new roots to grow.

In general, it's safe to remove up to half of the old soil when you repot.

If you're repotting a tree for the first time, it can be worth investigating to find out what the soil is like in different areas of the root ball. If you find a lot of old, compacted soil, try creating a few large holes in the root ball with a chopstick or removing wedge-shaped sections to introduce new soil into otherwise hard-to-reach areas. You can replace the remainder of the old soil the next time you repot in two to three years.

For fast-growing deciduous species such as Japanese or trident maples, it's generally okay to bare-root healthy specimens by removing all of the soil. Bare rooting is not recommended for conifers.

Perforating the root ball

Root work complete

4. Secure the tree in the container.

   Put a layer of coarse soil on the bottom of the pot. Using a coarser mix here prevents the lower portion of the root ball from staying too wet.

   Add a layer of bonsai soil. Use enough to form a mound onto which you'll place the tree in order to prevent air pockets forming under the root ball.

Drainage layer

Adding soil to the pot

A mound of bonsai soil

Place the tree onto the mound of soil and twist it back and forth as you press down until the tree is set at the appropriate front, potting angle, and potting height.

Once the tree is in place, secure tie-down wires to hold the tree in the pot (see "Securing Bonsai in the Pot," page 120).

Setting the tree

Securing the tie-down wires

5. Fill the container with soil.

   Use a chopstick or slender piece of bamboo to work the soil between the roots with a deliberate jabbing or poking motion. Doing this helps stabilize the tree in its container and eliminates large air pockets that can lead to root dieback.

   Start at the front of the tree and work your away around the root ball without working the same area more than once.

   Press firmly enough so that the chopstick goes down into the loose soil but not so firmly that it disturbs the drainage layer. As the soil firms up, work on shallower and shallower layers of the soil until you're working near the surface. Take care to avoid crushing akadama particles (a sign you're working too forcefully).

   Add soil, if needed, to bring the level just below the lip of the pot.

Working the soil into the root ball with a bamboo stick

Adding soil

Use a hand broom at a low angle to smooth the surface of the soil.

Tap the surface of the soil with your hand or a small trowel to lock the soil particles into place. (If the surface soil moves around when you water, roots won't grow in it.)

6. Water the tree.

   Gently water the tree until the water coming out of the drainage holes is clear to rinse away any dust or fine soil particles in the pot. (This may take several minutes.)

   Use a watering can or hose with a fine nozzle to prevent soil particles from moving around.

Smoothing the surface of the soil

Tapping the surface of the soil

Repotting complete

Securing Bonsai in the Pot

Although there are many ways to secure a tree into its container, the goals are always the same: to prevent the tree from falling out of the pot and to protect the roots from damage caused by inadvertent movement of the trunk. Depending on how many holes a pot has or how the holes are arranged, you can use different approaches for each configuration.

FOUR-HOLE POTS

If your pot has four holes that can accommodate tie-down wires, you can secure a tree using a four-hole tie, also known as a box tie.

1. Cut two pieces of wire the length of two long sides of the pot and one short side.
2. Pre-bend the wire to 90 degrees in the shape of a large "U."
3. Insert the tie-down wires.

   When the tie-down wires are in place, add soil to the container and set the tree.

1. Measuring the wire

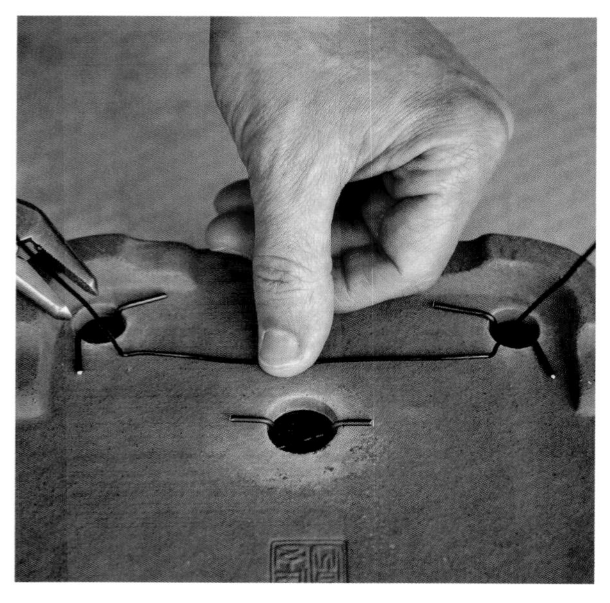

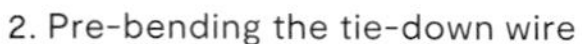

2. Pre-bending the tie-down wire

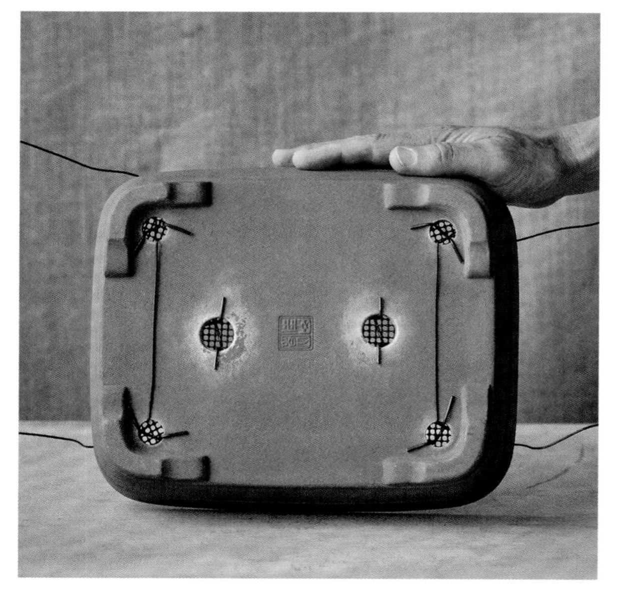

3. After inserting the tie-down wires

4. Create a "tail" by making a loop at the end of a small wire and thread the first tie-down wire through it.
5. Pull the wire with the tail over the front edge of the root ball toward the next wire and twist them together.
6. Repeat the process with the next wire.

4. Tie-down wire inserted through loop at end of tail

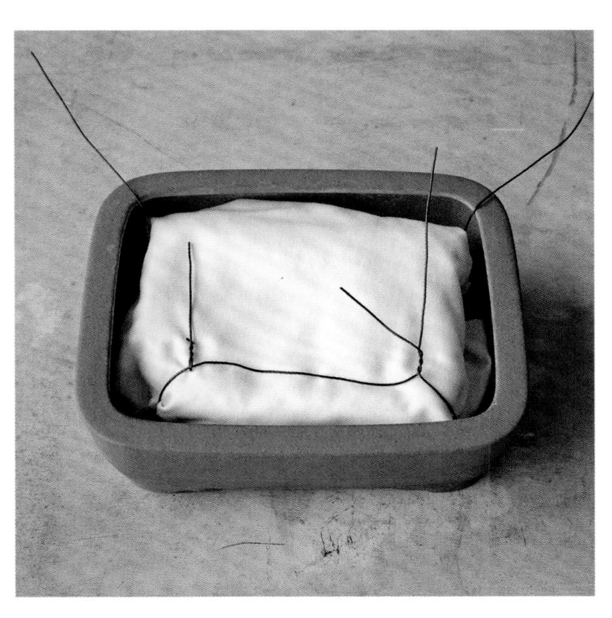

5. Connecting the first two wires

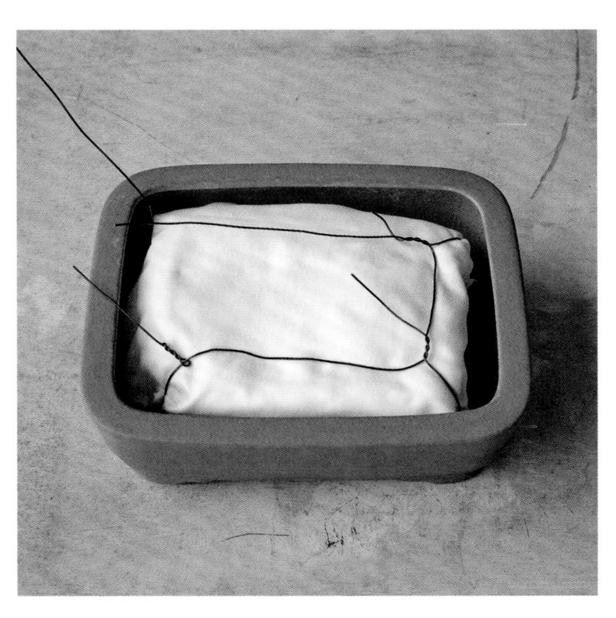

6. Connecting the second and third wires

7. Repeat again.
8. Connect the last open wire to the tail.
9. Pull wires taught with pliers and slightly release tension as you twist. Cut the ends of the wires.

If the tree you are repotting leans strongly to one side, ensure that the final connection (the point where you connect the last remaining wire to the tail) is on the opposite side of the lean. In other words, if the tree leans to the right, make sure the final tie is on the left side.

7. Connecting the third and fourth wires

8. Connecting the fourth wire to the tail

9. After tightening the final connection and cutting the ends of the wires

ONE-HOLE POTS

1. Cut two tie-down wires equal in length to the circumference of the pot. Create a "cotter pin" and wrap the two tie-down wires around it.
2. Insert the tie-down wires into the drainage hole. The cotter pin will hold the wires in place.
3. Connect the wires in pairs and secure them with pliers.

Because one-hole pots provide the least amount of stability when securing your trees, consider drilling additional holes when possible so you can use the techniques below for two- or three-hole pots.

1. Bobby pin with tie-down wires

2. Prepared one-hole pot from below

3. After securing the tie-down wires

TWO-HOLE POTS

1. Insert two tie-down wires.
2. If the root ball is solid, you can secure the tree with chopsticks or slender pieces of bamboo. Insert the chopstick into the root ball with your hands or drive it in with a small hammer.
3. Connect the ends of the wires over the root ball (or over the chopsticks) and secure them with pliers.

1. Prepared two-hole pot from below

2. Securing the root ball with a bamboo chopstick and wire

3. After securing the tie-down wires

THREE-HOLE POTS

1. Insert two tie-down wires into the drainage holes.
2. Connect the wires as with a two-hole pot.
3. Connect the tails to each other and secure with pliers.

As with the four-hole pot, ensure that the final tie is opposite the direction in which the tree leans.

1. Prepared three-hole pot from below

2. After connecting the first two wires

3. After connecting the tails

Repotting Aftercare

In most cases, you can return trees to their usual place in the garden after repotting. Strong wind can dry trees quickly, so avoid keeping recently repotted bonsai in exposed areas. Cold-hardy species can handle brief overnight lows at or just below freezing, but don't hesitate to protect weak trees, small trees, and precious trees as temperatures approach freezing. When temperatures drop lower, it's best to shelter recently repotted trees in a workshop, garage, or heated greenhouse.

Water recently repotted bonsai regularly (about once a day when it's sunny and dry) to prevent new roots from drying out. If this is the first time your tree is growing in bonsai soil, you may have to water more frequently than before. Watch watering needs carefully until you become familiar with how your tree responds to the new soil.

Signs of a Successful Repotting

You can say you have successfully repotted a bonsai when you get the following right:

- You're happy with how the tree is positioned in its container.
- The tree is well secured in the container and resists gentle attempts to rock it back and forth.
- The container is well suited to the tree.
- The soil you selected supports healthy root growth and overall tree health.
- You can water the tree easily.
- The tree grows well over the next growing season.

Deformed leaves, stunted growth, and weakness—whether general or localized to one or more branches—are signs that the tree hasn't recovered from repotting. If you see any of these signs, avoid pinching, pruning, or wiring until the tree is growing normally.

It's common to delay bonsai work for one year after an aggressive repot, but it's fine to work on young trees that grow well after repotting. The exact amount of work you can do will depend on the species, the health of the tree, and the extent of the root work. Checking for signs of stress before working on a tree that's been repotted and planning your work accordingly is the best way to maintain the health of recently repotted trees.

MICHAEL HAGEDORN

Finding the right container for a given tree can be tricky. If the two aren't a good match, the container may overwhelm the bonsai, making the trunk look too slender or the design look awkward. This is not lost on Michael Hagedorn. Michael was a full-time bonsai potter before he traveled to Obuse, Japan, to study bonsai with Shinji Suzuki.

When Michael returned to the United States, he found himself working with species and forms that differed from the trees he worked with in Japan. Which containers did he think best complemented these trees? Often, the answer was no container at all.

Michael has been experimenting with growing bonsai on slabs and using nothing but moss and "muck" (see below) to keep the soil from falling away. These compositions feature mountain hemlock, vine maple, and ponderosa pine, among other species, and are now among his best-known works.

Michael notes that the bonsai container traditionally provides contrast to the tree: "It's an inorganic element that, when combined with an organic element, lets us add meaning to a composition by highlighting or downplaying different aspects of the tree such as the bark, the foliage, or the deadwood."

Michael has a fondness for less "energetic" trees such as literati, clump style, or other bonsai forms featuring slender trunks, and he finds they are particularly well-suited for the "softer" feeling that a moss-covered slab can convey. "Removing the container removes a key aspect of the editorial role bonsai practitioners play when they select a growing environment for their trees," he says. I couldn't have said it better!

If you're interested in trying this approach, here are some tips from Michael for growing bonsai without containers:

- Use trees with established root balls. It's easier to carve an existing root ball into the shape you want than to start with a root ball that lacks integrity and build up a structure with loose soil.
- Use a solid base. Materials such as stone, wood, cutting boards, or even solid-surface countertops can be used as "containers" for the tree.
- Build up the walls that will hold the soil in place with muck. Michael's simple recipe is a blend of equal parts cornstarch, long fiber moss, and akadama fines. First, add the cornstarch to water and stir. Cook the mixture in a microwave or on the stovetop until the texture becomes like jelly. When the mixture is cool enough to handle safely, mix it with the long fiber moss and akadama fines.

CHAPTER 7

Summer and Winter Care

Bonsai that are happy in their environment, well suited to the climate they grow in, and adequately protected from temperature extremes in summer and winter, can develop quickly. By using the strategies discussed in this chapter, you can learn to make the most of the growing conditions in your garden.

ZONE ENVY AND CLIMATIC LIMITS

Chances are good you have a few trees in your collection that aren't native to the area in which you live. That's to be expected. After all, part of the appeal of growing bonsai is that it provides you an opportunity to work with trees that suggest climates, regions, and landscapes that differ from your own.

To determine if a given species is a good fit for your climate, give it a try with a tree that's not too precious. If it produces vigorous growth and gets healthier every year, it's likely a great fit for your garden. If, however, the tree you're testing becomes weaker over time or simply fails to thrive, the species may not be a good match. When this happens, you may experience "zone envy," the feeling you get when you want to grow species that aren't well suited to your climate.

I know this feeling well. Over the years, I've tried to grow many species that just aren't equipped to thrive in my region (mountain hemlock, for example). I thought that if my horticulture were good enough, I might be able to tip the scales, but to date I've found this to be wishful thinking.

If you want to get ideas for species that grow well in your area, start by checking with local bonsai clubs or enthusiasts. Local nurseries may also be able to help. If you don't have local resources, online bonsai communities are a natural next step.

You can also look around and take note of what grows well in your area. If it grows like a weed, it could be a great candidate for training as bonsai. The same principle can apply to trees already in your collection. Even when a species is theoretically a good match for your climate, you may find that it's more likely to get sick, attract insects, or recover more slowly after you work on it than other species. If this sounds familiar, refocusing your collection on species that thrive can make bonsai a much more satisfying activity.

ORGANIZING YOUR GARDEN

How you arrange the trees in your garden can have a huge effect on their health. Start by providing your trees with the appropriate amount of sunlight for each species. Filtered light is good for deciduous trees such as maples. Full sun is best for most conifers and heat-loving species such as oaks, bougainvillea, and ficus.

Move trees to shadier areas of the garden if you see signs of sunburn (pale or brown patches on leaves that face the sun) or if trees appear stressed in summer (overly compact foliage, little new growth, pale color). Signs of too little light include leaves that are larger or darker than normal or a lack of new growth. Once you've determined the best exposure for your trees, don't hesitate to move them around throughout the year to take advantage of changing light conditions.

When placing trees on benches, be sure to leave ample space between them. Crowded benches can reduce airflow, make watering difficult, and lead to unbalanced growth. Leaving ample space between trees also makes it easy to rotate them periodically (at least once a month) to ensure that all sides of the tree receive equal light over the course of the year.

You can also group trees with similar water needs to simplify watering for yourself and for whomever waters when you're not around. This may mean grouping trees by type (conifer versus deciduous), size (placing small trees together to facilitate frequent watering), or species (grouping trees that don't require frequent watering can prevent overwatering).

A similar strategy can work to reduce conditions that lead to pest and disease problems. If, for example, you regularly see powdery mildew on your trees, move them farther apart, increase the sunlight they receive, reduce sources of excess moisture in the garden (such as landscape material or old leaves), and avoid overhead watering or watering late in the day.

Alternatively, if it's hot and dry where you live, consider planting groundcovers, shrubs, and trees in the garden that can serve as windbreaks and increase humidity around your trees.

Pests and diseases can spread quickly, so acting when you first notice symptoms will make it much easier to keep problems under control. If you find pests are a recurring problem, schedule periodic checks on your trees to identify problems as soon as possible.

Caterpillar on cork oak foliage

SUMMER CARE

Dried moss, white sphagnum moss, and a sifted mixture of the two

Running moss through a sieve

Sphagnum moss top dressing on a Japanese maple

Providing bonsai with the right amount of sunlight during the growing season can mean very different things, depending on where you live. In milder climates, you may strive to provide as much sunlight to your bonsai as possible. In warmer climates, your goal may be to mitigate heat events that can last for days or sometimes weeks. Here are some tips for keeping trees healthy in summer.

Top Dressings

An easy way to protect trees in summer is to apply a top dressing. Because soil near the surface dries out before the rest of the root ball, roots are often reluctant to grow in the top part of the pot, especially in dry climates. Top dressings are particularly useful for deciduous species, as they facilitate root growth that can improve the *nebari*.

You can make a top dressing by running white sphagnum moss (also known as orchid moss or New Zealand sphagnum moss) through a sieve. Use different sized sieves depending on how coarse you want the top dressing to be. Mixing live or dried moss with the sphagnum can encourage a layer of moss to grow in a matter of weeks if the surface of the soil stays moist.

Humidity Trays

If your small trees or accent plants regularly dry out before you have a chance to water, a humidity tray can help. To create one, fill a shallow container with pumice, scoria, or similar particles and nestle your bonsai into the material in the tray. Humidity trays work by providing additional moisture and insulation for the roots. The deeper you nestle the plants into the tray, the better you can insulate their roots.

If your humidity tray has no drainage holes, you can fill it with water and place the pot on top of the soil in the tray so the roots don't become waterlogged. Trays with drainage holes, on the other hand, can serve as extensions of the pot. When bonsai grow roots into the soil in the tray, they are better able to withstand hot weather without suffering damage.

Humidity tray

Bonsai garden with summer shade structure

Shade Cloth

Depending on your climate and the types of trees you grow, you may find it challenging to keep them all healthy in full sun. Some signs that your trees are receiving too much sunlight include yellow foliage, sunburn, and frequent drying or wilting. If your garden doesn't offer the right amount of protection in summer and you're noticing these symptoms, adding shade cloth can help.

Shade cloth is a type of fabric designed to filter out a specific percentage of light. For example, 30 percent shade cloth filters out 30 percent of the sunlight. In mild climates where summer highs are between 80° and 90°F, you may find that 30 percent shade cloth is adequate to protect deciduous trees in summer. When summer highs reach 90°F and above, 40 or 50 percent shade cloth may be more appropriate. Results with shade cloth can vary widely depending on the species you grow and the growing conditions in your garden, so experience will be your best guide for providing your trees with the right amount of protective shade.

Many shade structures are seasonal. A common approach is to put up shade cloth when temperatures rise in summer and take it down in fall when the threat of heat waves passes. The best systems can be used on a day-by-day basis to give trees more light on cloudy or rainy days and less light when the sun is shining brightly.

TIPS FOR USING SHADE CLOTH

— Different species have different light requirements so you may find it helpful to set up shade structures that block different levels of light. For example, you can place the most shade-loving species under 40 or 50 percent shade cloth and species that can handle more light under 30 percent shade cloth.

— If your garden is exposed to drying winds, install shade cloth on the sides of your structure to offer additional protection and preserve humidity.

— Keep an extra sheet or two of shade cloth on hand. If you need to cover more of your garden during a heat wave, you can use the extra material to protect your trees.

GREAT TREES FOR HOT CLIMATES

— Bougainvillea
— Chinese elm
— Crepe myrtle
— Ficus
— Ginkgo
— Hackberry
— Oak
— Olive
— Pomegranate
— Silverberry
— Styrax

Heat Waves

When temperatures reach ten or twenty degrees above normal summer highs, bonsai require extra protection to avoid sunburn or heat stress. In general, the higher the temperature above normal, the more care is required to keep your trees healthy.

For starters, avoid working on trees if you know a heat wave is coming. Wait until the heat wave is over before wiring, pruning, grafting, fertilizing, or spraying your trees with pesticides and fungicides.

The next thing to do is to provide additional shelter for sensitive trees.

- Move trees out of the sun to a shady part of the garden or under benches to provide shade and reduce wind exposure.
- Move trees away from fences, walls, or other structures that reflect light and heat, which can increase the risk of sunburn.
- Set up temporary shade structures (or double up the shade cloth on existing structures) to reduce the sun's intensity and protect your trees from drying winds.
- If there is no good shelter outside, move trees into a workshop, garage, or even into the house to reduce exposure to extreme temperatures.

The most important trees to move are those most likely to suffer from excessive heat, including small trees; rock plantings; trees that were recently pruned, wired, grafted, or repotted; cuttings and seedlings; sick trees (any trees suffering from pests, disease, or general weakness); and precious or old trees that you don't want to risk damaging. You can protect trees that stay outside by wrapping the pots with towels, burlap, or other materials that preserve moisture and shielding the containers from direct sunlight.

After providing shelter to the trees that need it, you can focus on watering.

Instead of following your normal watering protocol during a heat wave, do the following:

- Water everything the day before the heat wave begins, to ensure that trees don't start the day dry.
- In addition to watering the roots, water the foliage and the outside of the pot to keep trees cool as needed throughout the day (see page 152 for more information about overhead watering).
- If the humidity is low, water everything in the area where bonsai grow, including benches, walkways, and fences, to cool down surfaces and increase the humidity around your trees. (If humidity is high or excessive watering causes fungus problems, don't water late in the day, and avoid getting the foliage wet.)

Take extra care for heat waves that last more than one day. Trees that may be okay in high temperatures for one or two days may show signs of stress when exposed to excessive heat for multiple days or weeks. Above all, watch out for drying winds during heat waves. If the humidity is low, wind on hot days can dry out trees quickly. Sheltering trees from sun and wind during heat waves will go a long way toward keeping your bonsai healthy.

WINTER CARE

Depending on where you live and what kind of trees you grow, your bonsai may need protection from the elements in winter. Cold, of course, is relative, and finding the best approach for your collection can take some time to figure out. Here's a guide for getting started.

Cold Hardiness

Preparing your trees for a successful winter begins with figuring out how much cold weather they can handle. One starting point is to check the USDA Plant Hardiness Zone Map to determine your "Plant Hardiness Zone" and then look up the zones in which your trees will grow to see if they are a match. Although hardiness zone designations can provide basic cold-hardiness information, they may not be the most accurate guide when it comes to winter care for bonsai. Because bonsai grow in containers, the roots get much colder than they would in the ground. This makes them more susceptible to damage at warmer temperatures than their hardiness designation suggests.

One of the best resources for identifying safe winter temperatures for your bonsai is the book *Bonsai Heresy* by Michael Hagedorn. In addition to providing an excellent guide to tree dormancy requirements, *Bonsai Heresy* notes the temperatures below which roots can experience damage.

Within a given species, you may find that some trees fare better in cold temperatures than others. This is because healthy trees can tolerate lower temperatures than sick trees, young trees can withstand lower temperatures than older and more refined trees, and trees with established root balls can withstand lower temperatures than trees that were recently repotted. You'll also find that trees growing in larger pots can withstand lower temperatures than trees in smaller pots. Considering these variables will help you provide adequate protection for all of your bonsai when temperatures drop.

Preparing Bonsai for Winter

Here are a few things you can do toward the end of the growing season to help keep your trees healthy through winter:

- Remove leaves from deciduous trees and brown needles from conifers.
- Treat for insects or disease before trees go dormant, as untreated problems can worsen over winter and leave trees weak in spring.
- Remove any fertilizer buildup on the surface of the soil. Organic fertilizers such as cottonseed meal can break down over the growing season and form a crust that makes watering difficult.

These tasks apply whether your winter is mild or severe. Once you've completed them, you can focus on making sure your trees receive the appropriate amount of protection for the season.

Overwintering Bonsai Outdoors in Mild Climates

If winters are mild where you live, you can likely keep your bonsai outdoors year-round. Brief or infrequent freezes during which temperatures drop no lower than 28°F for a few hours are fine for most non-tropical species including deciduous trees such as maple, beech, and elm, as well as conifers such as pine, spruce, and juniper. Many broadleaf evergreens such as oak, olive, or azalea may be able to handle brief freezes, but protecting these species from freezing temperatures can preserve the color of the foliage and prevent branch dieback.

Most tropical species that grow outside will need protection when temperatures drop below 50°F. If you have a small number of tropical bonsai such as ficus or jade growing outdoors, the easiest solution may be to bring them indoors during the winter months. Place these trees in a bright, warm spot. Continue to water when the soil begins to dry out, and rotate trees regularly to make the most of the available sun exposure. If you don't have space inside, consider placing tropical bonsai in a heated patio, sunroom, or greenhouse kept at around 70°F.

Overwintering Bonsai Outdoors in Cold Climates

If freezes are common where you live and temperatures regularly drop below 28°F, your trees may require additional protection in winter. When a cold spell is coming, move sensitive trees under benches to capture heat from the ground and shield them from wind. If there's a top dressing such as gravel or mulch on the ground, nestle each pot a few inches into the top dressing for additional protection.

Burying trees in the garden up to the lip of the pot or placing them in mulch beds during winter can provide more insulation for the roots, but this approach lacks the flexibility you get when keeping trees in cold frames or greenhouses. If you do store trees on the ground over winter, find a shady location next to a fence or wall to cut down on wind and reduce temperature fluctuations. When possible, shovel snow up to the first branches to provide additional insulation.

Tree on a block to reduce moisture in the soil

Excessive rain can also be stressful for bonsai, particularly for sick or weak trees. If your area receives constant rain or experiences an especially wet year, place a wood block under one side of the pot of any weak trees to reduce the amount of water retained by the soil. If conditions remain wet, give weak trees a break by placing them under an eave or moving them out of the elements so the soil in the pot can begin to dry out.

Overwintering Bonsai Indoors

If your climate requires that you keep your trees indoors over winter, you can place them in a structure created for plants (such as a greenhouse or cold frame) or repurpose an existing space such as a shed, basement, enclosed patio, or garage. By far, a greenhouse offers the most flexibility for protecting your trees during winter, but it also requires the most experience to make sure it works the way you want it to.

Experienced greenhouse growers will tell you that it takes time to build up the expertise required to maintain safe temperatures for bonsai over winter, so don't expect to get everything right the first time (or even the second!).

Wintering bonsai indoors with grow lights

If you're thinking of setting up a greenhouse or cold frame for overwintering your trees, take steps to make sure it doesn't get too hot on warm or sunny days. If necessary, cover it with shade cloth (or hang shade cloth from the inside if you expect significant snow loads). Temperature-controlled exhaust vents can also help maintain cool temperatures on warm days, as can opening doors and windows. Fans (to increase air circulation and reduce fungus problems), heat pads or heat coils (if root dieback is common), and supplemental lights (if trees turn yellow or grow weak because of low light conditions) can make greenhouses, or any indoor storage areas, more hospitable for bonsai in winter.

Trees respond to cues such as day length and temperature to initiate different aspects of dormancy. To help kick off this process, make sure bonsai experience some cold weather before bringing them indoors for the season. Once trees are tucked away for winter, the goal for many species is to maintain temperatures between 33° and 40°F, which are cold enough to meet dormancy requirements without exposing trees to harmful lows.

If you're new to overwintering bonsai and have questions about what's best for your trees, a good starting point is to consult with people who grow similar species in similar climates and experiment from there.

TIPS FOR WINTER CARE

- **Keep trees well-watered.** Dormant trees don't completely shut down in winter, so they'll need regular watering about once a week. Water more frequently if your trees dry out or start growing.
- **Protect trees from animals.** Rats, mice, voles, deer, gophers, and rabbits have been known to eat or otherwise damage bonsai over winter.
- **Protect delicate pots.** All but the toughest bonsai containers are susceptible to breaking during hard freezes.
- **Reduce temperature fluctuations and freeze-thaw cycles.** Trees that freeze and thaw frequently experience more stress than trees that remain just above or just below freezing through winter.

Greenhouse for protecting bonsai in winter

Removing Trees from Winter Storage

As the days lengthen and temperatures warm up, trees in winter storage will begin to grow. When this happens, move bonsai into the sun to prevent new growth from becoming weak or leggy. If, however, nighttime temperatures drop below freezing or cold weather returns, move the trees back to protected areas at night until the weather gets warmer.

Once your trees have been outside for a month or two, you can begin to gauge the success of your overwintering strategies. If your trees are healthy and growing well, you'll know you're on the right track. If your trees are weak in spring or they lose branches as they come out of dormancy, consider providing more protection the following winter. Whatever your approach, over time you'll learn what works best for your trees.

GREAT TREES FOR COLD CLIMATES

- Amur maple
- Beech
- Birch
- Eastern red cedar
- Eastern white cedar
- Jack pine
- Japanese maple
- Larch
- Plum (cold-hardy species)
- Ponderosa pine
- Red maple
- Rocky Mountain juniper
- Spruce
- Sugar maple
- White pine

CHAPTER 8

Watering and Fertilizing

Although our trees can't talk, they can do a pretty good job of providing clues about whether or not they're happy with the water and fertilizer we provide for them. This chapter will help you better understand these clues so you can water and fertilize effectively.

WATERING BONSAI

Bonsai care begins with watering. This daily practice defines the bonsai lifestyle and forms the core of your relationship with your trees. When you're learning to water bonsai, keep in mind that watering too frequently is just as problematic as letting them get too dry. In addition to providing moisture for the roots, watering forces stale air out of the root ball and draws in the fresh air roots need to be healthy. As time passes between waterings, the amount of moisture in the soil steadily decreases as water is removed by the roots and by evaporation. When you water your trees, your job is to make sure that you let enough time pass between waterings to prevent the roots from staying too wet, but not so much time that the roots don't have access to the water they need to grow.

How to Water

When watering bonsai, your goal is to saturate the root ball completely. There's no need to worry about giving trees too much water at any one time, as any excess will run out through the drainage holes in the bottom of the pot. Use a watering can or hose-end nozzle with a gentle spray to avoid washing soil out of the container. If soil particles move around when you water, tamp loose soil into place and try to water more gently next time, as roots won't develop in loose soil.

Because water doesn't always flow through soil quickly, you may have to fill the container several times before the root ball is saturated. A good way to do this is to water your trees once, give the water an opportunity to soak in, and then water a second time. You can repeat as needed until you see water running out of the bottom of the pot.

If your trees dry out faster than expected, this can be caused by water not percolating through the root ball evenly when you water. You can investigate by checking the soil just below the surface. If you find dry soil, the following strategies can help.

- Remove the top ½ inch to 1 inch of soil with a chopstick or bent-nose tweezers and replace it with fresh soil.
- Perforate the root ball (with an awl, pick, or screwdriver) to improve percolation.
- Periodically place the tree in a water basin (pot and all) and let it soak for up to thirty minutes to give the root ball time to absorb water when the soil is old or compacted.

You can also use a combination of these approaches to ensure that the roots receive adequate water. Keep track of trees that require these treatments, as they are good candidates for repotting the following spring.

OVERHEAD WATERING

Your trees will appreciate it if you water the foliage when you water the soil in the following circumstances:

— When temperatures are high

— When foliage is dusty

— When spider mites are a problem

Most conifers can benefit from periodic overhead watering in summer. You don't need to water the foliage on conifers daily, but it can be a good weekly practice. Avoid overhead watering when diseases like needle cast or tip blight are a problem, when you want to preserve flaky bark or deadwood features that are susceptible to decay, or when you want to discourage moss from growing on the trunk or branches. Overhead watering can benefit broadleaf evergreen and deciduous species too, but avoid overhead watering trees that are susceptible to fungal diseases such as powdery mildew or leaf spot.

Determining Water Needs

The best way to determine whether or not bonsai need watering is to check the moisture level just below the surface of the soil. You can gauge dryness visually or with your fingers. Avoid using moisture meters as the results may be inconsistent due to the composition and particle size of common bonsai soil ingredients.

For bonsai with a moss top dressing, you can feel the surface of the moss with your hand to determine its dryness. For trees that prefer moist environments (such as most deciduous species), water when the moss starts to feel dry. For trees that prefer dry environments (such as pines), water when the moss feels completely dry to the touch.

As a baseline, water bonsai daily. If conditions are wet and the soil is saturated, you can skip watering for the day. If you find your trees drying out between waterings, you'll need to water more frequently.

Because your trees' water needs can change day-to-day depending on the weather, and over the course of the year as the seasons change, checking the soil regularly is the most effective way to make sure you're watering appropriately. A simple strategy for success is to schedule daily watering checks. When the weather is mild, you can schedule one watering check per day, ideally in the morning or midday, during which you water any trees that are dry and skip the rest. When temperatures increase to the point at which your trees need watering more than once per day, schedule two (or more) watering checks each day.

If you have trouble keeping trees from drying out or your non-bonsai obligations prevent you from watering when your trees need it, you can adjust your soil mix to support your watering habits. Using smaller particles or increasing the percentage of particles that retain more moisture (such as akadama) in your soil mix can reduce the rate at which the soil dries out. The same approach can work if the soil stays too wet. By using larger particles or soil components that retain less water (such as pumice), you can accelerate the rate at which the soil dries out. The more consistently the soil dries out from tree to tree in your garden, the easier it will be for you to make sure that all of your trees receive the appropriate amount of water.

Moist soil just below the surface; a good time to water deciduous or water-loving species. For trees that prefer drier environments, wait until the soil below the surface begins to dry.

Tips for Watering Effectively

Over time, ineffective watering habits can produce symptoms in your trees that let you know something's wrong. In some cases, the symptoms make it clear whether the trees are receiving too much or not enough water.

SIGNS OF UNDERWATERING

- Leaves wilt when the soil is dry and improve when the tree is watered.
- Leaf tips turn brown suddenly when the soil is dry.

SIGNS OF OVERWATERING

- Leaves appear wilted even when the soil is wet.
- Leaf tips gradually turn brown or develop spots.
- Soil smells bad (particularly during repotting).

Other symptoms such as yellow foliage and general weakness can be caused by overwatering *or* underwatering. In such cases, your best guide is your own sense of whether your trees are staying too wet or too dry. Consider which species grow better in your garden: those that need more water (such as ume or azalea) or those that need less water (such as ponderosa pine). And when you water, notice whether the soil is bone dry or still wet from the previous watering.

Experienced bonsai growers don't rely on special techniques to determine whether or not bonsai need water, but they do have a good sense of how growing conditions can affect a tree's water needs.

Water needs increase when

- the weather is warm, sunny, windy, or dry
- a tree is growing in full sun, has a lot of foliage, or has a sacrifice branch
- a tree is growing in a shallow container or the pot is full of roots

Water needs decrease when

- the weather is wet, cool, or cloudy
- the tree is sick, dormant, growing under shade cloth, or has little foliage
- the tree is growing in a deep container or has been recently repotted

Over time, you'll find that learning to take cues from your trees, and from the weather, can help you adjust your watering habits so your trees stay healthy.

Dry tips on maple leaves—a sign that the tree dried out or is underwatered

Wilted shoot on a cork oak—a sign that the tree needs water

Salt damage on Japanese maple foliage—a sign of poor soil or water quality

Healthy pine foliage and foliage with yellowing at the base of the needles—a sign of overwatering

WATER QUALITY BASICS

Do you know if your water is suitable for growing bonsai? Water quality can vary widely, whether it comes from a tap or a well, and can cause or contribute to a variety of symptoms in your trees. Among the most common symptoms are general weakness, discolored foliage, and leaf burn, particularly at the leaf tips or margins (the outer perimeter of the leaf). In more severe cases, poor water quality can cause leaf drop, twig dieback, and slow, stunted, or deformed growth. Less obvious signs of poor water quality include decreasing health from year to year or an increased susceptibility to insects or disease. If you see any of these symptoms on your trees and suspect your water quality may play a role, your next step is to learn more about your water.

Understanding Your Water Quality

Municipal water often contains minerals such as calcium and magnesium, among others, that affect its taste and smell. These minerals can be beneficial for plants in the right amounts or detrimental to their health when levels are too high or too low. You can get an overview of what's in your water by checking your local water quality report. You can usually find this online by typing the name of your water municipality followed by the words *water quality report* or *water analysis*.

Before getting into the details about how to evaluate your water, keep in mind that not everyone needs to check, let alone treat, their water. If you don't see persistant problems in your garden and live in an area that generally has good water, you may be in the clear.

TOTAL DISSOLVED SOLIDS

The first item to check in your water quality report is a measure of the total dissolved solids (TDS). The TDS refers to the sum of all organic and inorganic materials that are dissolved in the water (everything that's not H_2O). TDS is measured in parts per million (ppm) or milligrams per liter (mg/L).

For perspective, at the upper end of the spectrum it's not unusual to find TDS readings between 300 and 500 ppm from municipal water in Los Angeles or Phoenix. In both of these cities, taking steps to reduce the dissolved solids in your water can make it easier to keep your trees healthy. At the other end of the spectrum, TDS readings closer to 50 ppm are the norm in municipal water from Seattle, Washington, and Charlotte, North Carolina. As a result, it's not typically necessary to treat the water in those cities.

Although overall TDS levels are good general indicators for determining whether your water can cause problems, relatively low TDS levels (under 150 ppm) aren't a guarantee that the water is suitable for your trees. If certain individual minerals (such as sodium) occur at relatively high levels (above 10 to 15 ppm), you may begin to notice symptoms in your trees, even when overall TDS levels are low.

pH

The next item to check on your water report is the pH. The water's pH is an indexed measure of its characteristics as an acid or a base. Water with low pH (readings below 7.0) is acidic. Water with high pH (readings above 7.0) is basic or alkaline. Neutral water (pure H_2O) has a pH of 7.0.

The pH of the root environment affects the rate at which plants absorb nutrients. Plants can absorb some nutrients (such as manganese and iron) better at low pH levels and other nutrients (such as calcium and magnesium) at higher pH levels. Although a few species grow well in alkaline environments (hackberry is an example), many species trained as bonsai prefer slightly acidic environments (pH 5.5 to 6.5), as most essential nutrients are generally available within this range. If your water's pH is too high (above 8 or 9) or too low (below 5), your trees may not be able to absorb some nutrients even if they're present in the soil.

Calcium deposits (the white residue) on pots can indicate high mineral content in water.

Addressing Water Quality Problems

If you suspect that improving water quality has a chance to improve the health of your trees, look for good TDS and pH meters. These devices (when properly calibrated and maintained) will help you establish a baseline for your water and monitor any changes you make to it.

If your goal is to reduce the dissolved solids in your water, you can take a few basic approaches. You can remove dissolved solids with filters or a reverse osmosis system, you can use an alternative source of water such as rainwater or groundwater, or you can blend clean or treated water with untreated water. If your goal is to lower the pH, you can reduce the overall solids or acidify the water.

pH meter

USING RAINWATER OR GROUNDWATER

Rainwater can be great for watering bonsai. If you can capture enough of it, you can use it to water your trees undiluted, or you can blend it with municipal or well water to improve its overall quality.

Groundwater (water from a well) typically requires testing before it can be used to water bonsai. A TDS meter will give you an idea of the overall dissolved solids, but it won't provide information about individual minerals. Sending a sample for testing at a lab is the best way to learn more about what's in your groundwater and inform how you fertilize.

REMOVING DISSOLVED SOLIDS

If you have clearly identified water problems that need addressing for your trees to be healthy, a reverse osmosis (RO) system can be an effective way to remove solids from your water. These systems work by forcing water through a semipermeable membrane that diverts dissolved materials into a waste stream (known as the brine or the concentrate) that is discarded.

You can expect an RO system to remove 95 to 99 percent of the solids in your water, which will result in very clean water with a nearly neutral pH. This water is so clean that you'll need to add in the nutrients your plants need to be healthy. There are two basic options for doing this: you can add balanced fertilizers to the water that include all the necessary nutrients, or you can blend the RO water with untreated water to take advantage of minerals that are present in the untreated water. Whether you add fertilizer or untreated water to the RO water, you can aim to deliver water to your trees with an overall TDS level of around 100 ppm.

Anyone who has used RO systems for watering bonsai will tell you that it's not always straightforward to come up with a perfect approach on the first try. That said, once RO systems are up and running, the water they produce can make it far easier to keep your trees healthy if your water quality is poor.

Inline hose filters are a simple alternative to RO systems and may be helpful if slightly lowering the dissolved solids in your water is enough to make a difference for your trees. If you decide to give hose filters a try, use a TDS meter to make sure it's working as expected.

LOWERING WATER pH

If you suspect the high pH of your water is affecting the health of your trees, acidifying the water may help. For small bonsai collections, you can add an acid used for irrigation water (such as citric acid) to the watering can when you water. For larger collections, you can use chemical injectors that draw acid from a reservoir and inject it directly into your water. In general, pH problems are more of a concern when TDS levels are high, so start by reducing the solids in your water before taking steps to adjust the pH.

MAKING CHANGES INCREMENTALLY

If you decide to treat your water, you may find that some changes show up quickly in your trees (such as new growth in summer) while other changes show up later (such as foliage that stays healthy through summer and drops later in fall). To the degree possible, try to make one improvement at a time, as changes to one part of the system can have unintended consequences on other parts of the system. And be sure to keep track of any changes you make so you can evaluate the results over time.

Chemical injector

FERTILIZING BONSAI

When trees receive the right amount of the right nutrients, they can grow vigorously, resist infestation and disease, and respond well to training techniques. When they receive too much or too little of a given nutrient, they're less likely to stay healthy or produce growth that supports your development goals. To help you fertilize effectively, we'll look at signs that trees need more (or less) fertilizer, provide starting points for fertilizing your trees, and note the pros and cons of different fertilizers so you can determine the best approach for your bonsai.

Fertilizer Basics

Fertilizer is any substance you apply to the soil to help plants grow. The main reason bonsai need fertilizer is that common soil mixes don't provide trees with the nutrients they need to produce healthy new growth and respond well to training techniques.

Fertilizer generally comes in two forms, liquid and solid, that determine how you apply it. A fertilizer's formulation can be categorized as either organic (made from plant or animal matter) or chemical (made from inorganic compounds).

From a technical perspective, fertilizer is not plant food. Plants make their own food in the form of sugars created by photosynthesis. To perform photosynthesis and produce healthy growth, however, plants rely on minerals known as essential plant elements or nutrients. The lack of any essential nutrient will limit plant growth even when plenty of other nutrients are available. As a result, the key to applying fertilizer effectively isn't to apply lots of everything, but to apply the right amounts of whatever is deficient in the soil.

The minerals plants need the most are called macronutrients. The first three, nitrogen (N), phosphorus (P), and potassium (K), are known as primary nutrients. The primary nutrients are listed on fertilizer labels as N-P-K. (A 3-2-1 fertilizer, for example, contains 3 percent nitrogen, 2 percent phosphorus, and 1 percent potassium by weight.) The other three macronutrients, or secondary nutrients, are calcium, magnesium, and sulfur. Bonsai need these, too, so look for fertilizers that provide all six macronutrients.

Minerals required by trees in smaller amounts are known as micronutrients or trace minerals. Micronutrients useful for bonsai include iron, chlorine, boron, manganese, zinc, copper, molybdenum, and nickel, among others. If your trees don't have access to adequate levels of micronutrients in the soil, you will need to provide them with fertilizer.

Signs of Underfertilizing and Overfertilizing

The most common signs of underfertilizing are general weakness, slow growth, and pale or discolored foliage. Underfertilizing can also increase susceptibility to infestation or disease.

The most common sign that a tree is receiving too much fertilizer is when it produces growth that is stronger than you want it to be to achieve your development goals. This can take the form of large internodes and thick branch tips on deciduous species, and the production of juvenile foliage on junipers. Overfertilizing can also lead to infestations of sucking pests such as scale insects or aphids.

A plant that doesn't have enough of a specific nutrient is said to have a mineral deficiency. Too much of a given nutrient can cause mineral toxicity (which is far less common than mineral deficiencies). The most common signs of mineral deficiencies are consistent patterns of discoloration, deformation, or dieback in the foliage, including the following:

- Dieback on leaf tips or margins
- New shoots that are stunted or malformed
- Yellowing that occurs between leaf veins, in leaf veins, at the base of the leaf near the petiole, or along the leaf margins
- The appearance of symptoms in new foliage but not old foliage (or the reverse).

Chronic dryness, overwatering, and pH problems can make it difficult for plant roots to absorb nutrients. Correcting these problems is the first step toward addressing nutrient imbalances. The next step is to apply fertilizers that contain a mix of macronutrients and micronutrients. For stubborn problems, the most accurate way to check nutrient levels is to send tissue samples to a lab for analysis. This is a good option for commercial growers, but may be impractical for bonsai hobbyists. An alternative is to consult with experienced nursery staff, plant researchers, or bonsai professionals to determine next steps.

Silverberry, Japanese maple, and pomegranate leaves with interveinal chlorosis (yellowing with green veins)—a common symptom of iron, zinc, and manganese deficiencies

Tips for Selecting and Applying Fertilizer

If you're relatively new to bonsai, select one or more fertilizers that provide the six macronutrients (and possibly some of the trace minerals) and then see how your trees respond. The results may not be immediate, but over time you'll get a sense of which products are a good fit for your garden. This may mean using a mix of products to make sure your trees get the nutrients they need to be healthy. Once you come up with a good fertilizer regimen, you can stick with it until your trees indicate they need something different.

Although it's more common for bonsai to be underfertilized than overfertilized (likely because of the effort required to meet trees' nutritional needs), that doesn't mean you need to use strong fertilizers. In general, using fertilizers with N-P-K levels in single digits (between 1 and 9 percent by weight) are potent enough to keep bonsai healthy if they're applied according to the directions on the label.

Tea bag filled with fertilizer

Liquid organic fertilizers such as fish emulsion and kelp, or dry products that combine a variety of organic ingredients, are popular starting points for fertilizing bonsai. These are less likely to cause fertilizer burn or mineral buildup over time, even if they're applied in excess.

In general, avoid fertilizing sick bonsai until you address any underlying causes of poor health such as insects, disease, or poor soil.

TIPS FOR APPLYING LIQUID FERTILIZERS

- Water trees lightly before fertilizing to help the fertilizer soak into the soil evenly.
- Water trees lightly after applying organic fertilizers such as fish emulsion to reduce the smell and rinse residue from the lip of the pot (but not so much that you wash the fertilizer out of the pot).
- Follow a schedule to ensure that trees get enough nutrition, as liquid fertilizers wash out of the soil during subsequent waterings.

TIPS FOR APPLYING SOLID FERTILIZERS

- Distribute clusters of organic fertilizer evenly around the tree about halfway between the trunk and the edge of the container. When it's time to make a second (or third) application, leave the first application in place, as some fertilizers release different nutrients at different rates.
- If the surface of the soil is sloped, use fertilizer covers (or tea bags filled with fertilizer and secured with toothpicks) to prevent fertilizer from washing away when you water.
- If solid fertilizer breaks down and impedes drainage, remove the top layer of soil and replace it with fresh soil.

Organic fertilizer on the surface of the soil

Fertilizer covers hold organic pellets in place

When to Apply Fertilizer

The best time to apply fertilizer is during the growing season, when trees are active. The precise start and stop times, however, depend on the type of tree and your goals for development.

For most bonsai, you can start fertilizing around the time you see buds begin to swell in early spring. If a tree has been repotted, wait at least four weeks, or until the tree starts growing, before fertilizing. You can continue fertilizing through fall when trees enter dormancy. If you grow bonsai in mild climates where trees don't go fully dormant, you can continue applying fertilizer through winter.

For mature trees or bonsai in the middle stages of development (when the goal is to build branch density), you may want to avoid applying fertilizer at times of the year when it can produce growth that's too vigorous. Fertilizing too much in spring, for example, can cause leaves and internodes to grow large on deciduous species or on pines that aren't decandled. Begin fertilizing these trees later in the season, after the spring growth has hardened off, to avoid causing unwanted elongation.

Pros and Cons of Different Fertilizers

When it comes to selecting fertilizer for your bonsai, you have a lot of options. A big part of your success will come down to how easy it is to apply and how likely you are to stick with a consistent regimen. Here are some pros and cons for using liquid and solid fertilizers, organic and chemical fertilizers, and chemical injectors.

Solid vs. Liquid Fertilizers

The choice between liquid and solid fertilizers comes down to convenience and personal preference. Trying both out to see which you prefer, or using a combination of the two over the course of the growing season, can help you determine a good approach for your trees.

SOLID FERTILIZERS

- Release nutrients slowly so you don't need to reapply them frequently
- Are visible on the surface of the soil, making it easy for you to see how much fertilizer each tree has received
- Break down over time and can clog the surface of the soil
- Release nutrients with every watering

LIQUID FERTILIZERS

- Release nutrients quickly but require frequent application because they wash out of the soil
- Have no visual impact on the garden (they don't clutter the surface of the soil), but there's no way to tell when you last fertilized or how much fertilizer a given tree has received
- Are less likely to clog the surface of the soil
- Wash out of the soil between applications

Organic Fertilizers vs. Chemical Fertilizers

As with the choice between liquid and solid fertilizers, the choice between organic and chemical fertilizers often comes down to how well a given fertilizer works in your garden. When evaluating the options, note that it may take more than one growing season before it becomes clear which approach provides the results you're looking for.

ORGANIC FERTILIZERS

- Release nutrients slowly
- Are not likely to cause fertilizer burn
- May support beneficial microbes in the soil
- May smell bad and attract animals

CHEMICAL FERTILIZERS

- Release nutrients quickly (time-release chemical fertilizers are an exception to this, as they release nutrients over time)
- Are more likely to cause fertilizer burn than organic fertilizers
- Are less likely to support beneficial microbes in the soil than organic fertilizers
- Don't typically smell bad or attract animals

Chemical Injectors

If you have a large collection, using a chemical injector can be a good way to apply liquid fertilizer to your trees. Chemical injectors work well with low-viscosity fertilizers (typically chemical fertilizers), but don't work as well with viscous fertilizers (such as fish emulsion).

Because chemical injectors work by adding fertilizer to your irrigation water, they make it easy to apply the same amount of fertilizer to all of your trees. Chemical injectors may not be as helpful if you have a small collection or apply fertilizers at different rates to different trees.

DAISAKU NOMOTO

Daisaku taught me the proper way to talk with trees.

I met Daisaku when he was an apprentice to Kihachiro Kamiya at the Kihachi-en bonsai garden in Anjō, Japan. Today he's an accomplished bonsai professional based in Miyazaki. He loves America and has visited many times. On his first trip to my garden, he gave me a tip that has stuck with me to this day. In response to a question about how to water, Daisaku said we have to ask the tree. Leaning toward a pine, he asked, "Tree, are you thirsty?"

Taking on the voice of the tree, he replied, "Yes, please give me water."

"Tree, are you hungry?"

"Yes, please give me fertilizer."

"Tree, are insects giving you trouble?"

"Yes, please help me get rid of them."

Our job, as he put it, is to learn how to interpret the signs bonsai provide so we can be better caregivers for them. If trees had voices, they could simply tell us what they need, but because they speak a different language, it's on us to learn how to communicate effectively with them. This conversation may not have stuck with me were it not for Daisaku's comically high-pitched tree voice, but the basic idea has helped me countless times over the years.

When I wanted to learn more about my trees' water needs, I spent a summer checking every tree in the garden daily to see which needed water and which didn't. Doing this took two to three hours a day, but it taught me more about my trees' water needs than years' worth of experience watering trees less carefully.

When I stopped following the fertilizer regimen I learned when I started bonsai and started using different fertilizers to see how my trees responded, I learned a lot about which products made the most sense for my garden.

When I started sending samples from sick trees to a plant pathology lab, I learned which pests and diseases to look out for and the most effective ways to deal with them.

In short, I learned to take my cues for bonsai care from the trees themselves, and I haven't regretted it once.

CHAPTER 9

Case Studies

Learning a new technique is only so helpful if you don't know how to apply it. The goal for this chapter is to help you you apply new techniques effectively by presenting case studies featuring a mix of species at different stages of development.

PRUNING TO MAINTAIN SIZE (JAPANESE MAPLE)

Shohin Japanese maple

After cutback

Visible buds

Multiple buds offer different points for cutback

One challenge with shohin bonsai is that they can grow out of shape quickly. This is especially true for fast-growing species like Japanese maple. To maintain this maple as a shohin bonsai requires regular pruning, at least once or twice per year.

Reducing about half of the branches will enable more light to reach the tree's interior, where it can encourage new buds to form and existing buds to open. Because the tree already has good branch structure, there is no need to shorten all the branches. (Bonsai in early stages of development or trees with flawed branch structures are better candidates for major cutback.)

When possible, cut back to visible buds. This is safer than pruning to a spot that doesn't have visible buds, since new growth doesn't always appear where you want it to. Thanks to regular thinning, this tree has lots of visible buds that provide good options for cutback.

GRAFTING TO IMPROVE LEGGY BRANCHES (SIERRA JUNIPER)

Sierra juniper—arrows indicate locations where scions will be grafted

The goal for developing this Sierra juniper is likely clear: to improve the branch structure and create a beautiful silhouette that complements the trunk.

Normally, the best way to improve the branch structure on conifers is to prune and wire the branches. In this case, however, the branches are too long and leggy to create a compelling silhouette. Some of this can be fixed by pruning. When a young shoot grows at the base of a long branch, you can remove the long branch and train the shoot to take its place. If, however, there is no foliage at the base of a branch, your next best option is to graft. Grafting is a great way to make bonsai more compact, as it enables you to place branches exactly where you need them.

Here are the basic steps for making a side-veneer graft. Late winter or early spring is the best time of year to do this work.

1. Select the scion (the small branch you will graft onto a new location). Healthy branches with elongating shoots are more likely to form a successful graft union than unhealthy shoots.
2. Wrap the scion with a permeable grafting tape (such as Buddy Tape) to prevent the scion from drying out while the graft union forms. Pre-stretching the tape facilitates application by softening it and making it self-adhesive.

1.

2.

3. Cut the base of the scion with a sharp grafting knife on two sides to create a wedge shape.
4. Determine where you want to place the graft so you know where to make the incision.
5. Insert the knife into the branch at an acute angle to create a flap. Make sure the knife cuts through the bark and into solid wood. Use a sliding motion by drawing the knife across the wood instead of pressing it straight through the grain. Avoid cutting more than halfway through the branch to prevent dieback.

3.

4.

5.

6. Insert the scion under the flap immediately to prevent sap from collecting where you cut. If the flap is wider than the scion, line up the scion with one side of the flap.
7. Secure the graft union (the spot where the scion has been inserted into the flap) with grafting tape.
8. If necessary, use cut paste to seal gaps in the tape to prevent the graft union from becoming waterlogged, as this can prevent the graft from being successful.

6.

7.

8.

AFTERCARE FOR GRAFTS

- If you live where the sunlight is intense, keep the grafted tree in a shady spot while the graft union forms and/or place a strip of masking tape on the top side of the scion (like a tent) to shelter it from direct sunlight.
- When the scion begins growing, gently unwind the end of the tape just enough to facilitate further elongation.
- When it's clear the scion has taken and is growing well, completely unwrap the foliage but leave the portion of the tape securing the scion to the host for about one year or until the grafting tape begins to cut in and cause unwanted swelling.

Dotted lines indicate the future silhouette

DEVELOPING BRANCH DENSITY (DWARF WISTERIA)

Dwarf wisteria

After pruning the new shoots

Spring growth before pruning, after pruning the new shoot, and after cutting the leaves

Developing wisteria bonsai is similar to developing other deciduous species. After the new leaves harden off in spring, you can cut branches back to the desired internode length and cut the leaves, if needed, to ensure that light reaches the tree's interior.

After the new shoots on this tree were shortened to about ½ inch, the foliage was still too dense for light to reach the tree's interior. Reducing each leaf to one or two pairs of leaflets maintains branch density while ensuring that the interior branches receive adequate light.

Healthy wisteria may produce new growth in summer—either new leaves and/or new shoots—depending on the overall health and vigor of the tree. If this summer growth is vigorous, the tree can be pruned (and the leaves can be cut) a second time in midsummer.

After cutting the leaves

PRUNING TO BALANCE VIGOR (CORK OAK)

This cork oak is in the middle stages of branch development. All of the primary branches are in place, but the lower branches need to thicken before the focus can turn to filling in the rest of the silhouette.

In terms of the tree's design, a wide trunk that quickly tapers to the apex suggests a full canopy with relatively wide lower branches. If these lower branches remain slender, they'll look out of scale with the rest of the tree. To make them look older, you can let them grow freely for several years until they reach the desired thickness.

Because cork oaks have strong apical dominance (the upper branches are naturally stronger than the lower branches), the lower branches won't get bigger unless the upper branches are kept in check with frequent pruning.

The present work for this tree is to shorten the upper branches and to position the lower branches with wire, but not cut them, so they can continue to thicken.

Before

After pruning and wiring—the dotted lines indicate the future silhouette

REMOVING SACRIFICE BRANCHES (COAST REDWOOD)

This coast redwood is in early stages of branch development. The current goal is to create fine branching on the upper half of the tree and large primary branches below.

The lowest branches have grown freely for the previous two years to help them thicken. Three of these four branches have reached the desired thickness so it's time to prune them to about 3 to 4 inches long. The fourth sacrifice branch is still relatively slender so it will thicken for one more year before it is reduced.

For the primary branches that were shortened, the focus over the course of the next year will be to begin developing the secondary branching.

Coast redwood

After shortening all but one of the sacrifice branches—the dotted lines indicate the future silhouette

MAINTAINING DEADWOOD FEATURES (SHIMPAKU JUNIPER)

Algae growing on the trunk

Removing the outer layers of bark

Brushing the bark

Cleaning deadwood with a toothbrush and water

Over time, deadwood features on bonsai can turn green and begin to rot. When you see these signs, it's time to clean the deadwood. Your weather, your watering habits, and the hardness of the deadwood will determine how frequently the jin and shari need cleaning. It's common to treat deadwood at least once a year to prevent it from rotting.

The first step is to remove the outer layers of bark on the lifeline. Doing this deepens the color of the bark and can reveal newly dead areas that can be treated as shari.

If desired, you can further brighten the color of the bark by brushing it with a soft wire brush or a stiff-bristled plastic brush.

After cleaning the lifeline, use a stiff-bristled brush and water to remove any algae or accumulated dirt on the deadwood. Use carving tools to remove rotten deadwood as necessary. If you need to preserve soft deadwood to maintain the tree's design, let the wood dry completely and apply a wood hardener.

Keep a spray bottle filled with water on hand to speed up the work. After brushing the deadwood, rinse off the tree to wash away any remaining debris.

Applying lime sulfur solution

After the lime sulfur dries

Before

After

Applying Lime Sulfur

A popular way to treat deadwood on bonsai is to apply a thin coat of lime sulfur. It's common to use 100 percent lime sulfur the first time you treat deadwood and lighter solutions for subsequent treatments. To avoid the bright white color that 100 percent lime sulfur treatments can produce, add one or more parts water for each part of lime sulfur. For darker solutions, add a drop of sumi ink.

The density of the wood will determine how much of the solution soaks in and how bright the resulting color will be. As a result, freshly carved deadwood may look different from aged deadwood on the same tree. Over time, as the deadwood ages, it will become easier to produce consistent results.

Apply lime sulfur when the deadwood is still damp but not wet. Applying lime sulfur to wet wood can lead to drips and streaks.

If you're planning to apply lime sulfur to a tree that needs pruning and wiring, it's common to do the deadwood work first. It's easier to clean jin and shari before wiring the branches, and discovering new deadwood features while you work may affect your design decisions.

STYLING BROADLEAF EVERGREENS (EUROPEAN OLIVE)

Olive before styling

After styling

Side view before styling

Side view after styling

Working with broadleaf evergreens such as the European olive (or with boxwood or silverberry) is similar to working with deciduous species. Although you have limited opportunities to bend branches larger than your thumb, you can expect new buds to form near the cut site when you remove or reduce large branches during the growing season.

Over time, you can use these new shoots to replace faulty branches (branches with uninteresting movement or poor structure) and increase branch density.

This olive has been trained as bonsai for a long time, but it hasn't been pruned or wired in recent years and has grown out of shape. Removing about half of the foliage on the tree will stimulate new growth to help fill in the silhouette and give the branch pads a greater appearance of age.

Over time, as additional shoots emerge along the secondary branches, you can continue to improve the silhouette by further reducing the upper branches and allowing the lower branches to elongate. This will give the tree a slightly more triangular shape that will make it easier to ensure that all of the branches receive adequate light.

STYLING CONIFERS (CRYPTOMERIA)

Cryptomeria, like coast redwood or Douglas fir, is known for producing straight trunks. To suggest the height of a full-sized tree, this cryptomeria has been styled as a far-view bonsai (it resembles a full-sized tree viewed from a distance).

Early summer is a good time to work on cryptomeria, as healthy specimens can produce a second flush of growth before entering dormancy in fall.

Reducing the foliage by about half will better define the branch pads and let light reach the lower branches. Wiring the branches sightly downward will suggest the silhouette of an older tree (branches on younger trees are more likely to grow upward).

Cryptomeria

After pruning and wiring

PRUNING DECIDUOUS SPECIES IN FALL (CHOJUBAI)

This tree is a dwarf cultivar of Japanese flowering quince known as chojubai. The species naturally takes a shrub-like form, which makes it a good candidate for making clump-style bonsai with many trunks and fine twigs.

Fall pruning typically begins when most of the foliage turns color and begins to drop. The first step is to remove all of the leaves and spent flowers, if any, so you can see the branches clearly.

After removing the leaves, you have a brief window of about one or two weeks in which to prune before the tree fully enters dormancy. This is a good time to remove flawed branches, improve the silhouette, and cut new shoots back to the desired internode length. Repeating this process for several years can create a full silhouette able to support lots of flowers when the tree blooms in early spring.

Chojubai in fall

After removing old leaves and flowers

After pruning

REFINING BRANCH PADS (WHITE PINE)

White pine

This white pine has been trained as a bonsai for a long time. The main development goal at this stage is to improve the silhouette by refining the branch pads.

A good time to work on white pines is toward the end of summer when the old needles turn brown. Before wiring, remove any old or discolored needles.

Since the tree has a relatively slender trunk compared with the silhouette, it's natural for the design to feature ample space between the branches. This makes it easy to see the structure of the tree, particularly the angles of the primary branches and the outlines of the branch pads.

When you're creating branch pads, the goal is to create a soft, rounded shape when the branches are viewed from the side, with even spaces between shoots.

Branch pads are typically larger near the bottom of the tree and smaller closer to the apex. At the crown, individual branches are often grouped together into larger blocks of foliage, as too many small blocks can look out of scale for a small tree.

Side view of a branch showing rounded pad shape

After wiring

BOON MANAKITIVIPART

Boon Manakitivipart kicked off my bonsai journey more than three decades ago. I can't overstate how much I've learned from him or how much he's contributed to the bonsai community.* If I could pick just one of his characteristics to highlight, it would be his passion for raising standards in bonsai.

Boon apprenticed in Central Japan with Yasuo Mitsuya and Kihachiro Kamiya. Struck by the beauty of the trees he was working with, he started thinking about ways he could raise the level of bonsai in the United States.

After wrapping up his apprenticeship, he started Bay Island Bonsai, an organization dedicated to producing an annual exhibit staged by the members. This goal provided focus to all the work we did throughout the year.

If the material we had selected to work on wasn't good enough for Boon, he told us to bring better material next time. If he found mistakes in our wiring, he told us to remove the wire and try again (and again) until we got it right. When it came time to stage the exhibit, Boon left it to the members to identify which trees received awards for their excellence.

By making students responsible for the quality of their trees, and for helping us recognize quality in one another's trees, Boon offered us valuable experience that's hard to come by when you're working on your trees alone. To this day, Boon remains a model for looking past what's "good enough" and for always encouraging us to do better.

* As a small example of his influence, Boon has worked with every professional featured in this book: Eric Schrader and I studied with Boon for many years, both Michael Hagedorn and Peter Tea worked with Boon before apprenticing in Japan, and Andrew Robson worked with Boon before apprenticing with Michael. Daisaku was Boon's senior during their apprenticeship with Kamiya.

CHAPTER 10

In-Depth Case Study

Up to this point, we've looked at different pieces of the puzzle: how to evaluate and design trees we find interesting, how to apply specific techniques, and how to keep trees healthy over time. In this last chapter, we take an in-depth look at the initial styling of a field-grown shimpaku juniper.

Due to the scope of the project, I spread out the work over the course of one year. This is a common approach when a tree requires significant work, in this case, carving, pruning, wiring, and repotting. The initial work took place in winter, with follow-up work occurring in summer and the remainder completed the following winter.

MAKING THE PLAN

My first step was to determine the best viewing angle of the trunk. After looking at all sides of the tree and testing out different angles, I selected the following front and planting angle. I found this view of the tree interesting because it showed off the curves in the trunk and created strong movement from left to right.

After determining the new planting angle, my next step was to determine the future silhouette of the tree. Here's how I came up with the plan. I wanted to create a design that highlighted my favorite part of the tree—the curves on the lower section of the trunk. Above that point, the trunk movement became less interesting, so I decided to convert this area into a deadwood feature. The branches on the right side of the tree would form the basic silhouette.

At a glance, the long runners at the top of the tree indicated that it was healthy and ready for styling. I began pruning the tree by removing about half of the foliage on the upper branches so I could convert them to deadwood features. My plan was to remove the remainder of these branches incrementally by pruning again in summer and in the following winter. By spreading the work out like this, I could give the tree time to respond to the initial work before removing additional branches.

Shimpaku juniper

Selected planting angle

After pruning

Back before pruning

Back after pruning

CREATING DEADWOOD FEATURES

Pen marks indicating the future jin and shari

Because deadwood features are prized in juniper bonsai, I decided to create shari along the trunk and make several jin near the top of the tree. I used a pen to indicate the location of the jin and shari to avoid accidentally removing too much of the lifeline. In general, the idea is to work with the grain when creating deadwood features, because cutting across the lifeline (cutting perpendicular to the grain of the wood) can result in branch loss when the connection between healthy roots and branches is broken.

I started the carving work by using a grafting knife to define the edges of the areas I wanted to carve.

To create the jin, I used concave cutters, root cutters, and pliers to peel away strips of wood. This approach allowed for the creation of deadwood features without leaving tool marks on the wood.

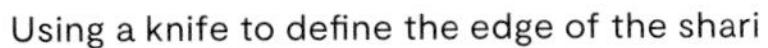

Using a knife to define the edge of the shari

After stripping away some of the bark

Starting at the end of the cut branch

Pulling away a portion of the branch

Using pliers to peel away strands of wood

Using a root cutter to further reduce the jin

Here's a close-up of the trunk after creating the initial deadwood features.

Close-up after carving

WIRING

Back branch to be moved

Making the bend

Back branch in its new position

The main goal for wiring at this stage was to position the primary branches and create a rough silhouette that I could refine over time. As this was a first styling, it wasn't necessary to wire all the fine branches.

The largest branch that needed moving was a back branch.

After applying the wire, I used both hands, one holding a pair of pliers, to move the branch into position. I was able to level the branch by twisting it counterclockwise while I moved it into place.

Moving this back branch made it visible on the left side of the trunk where it added depth to the design and created a more balanced silhouette.

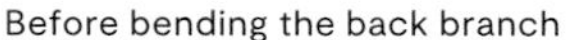

Before bending the back branch

Back branch visible from the front after bending it into position

I wired the rest of the branches to establish the basic silhouette.

Front

Back

Close-up of the future silhouette

Although I decided against repotting the tree to give it time to respond to the work, I shortened some of the longest roots to stimulate new root growth closer to the trunk. I also indicated a few additional roots to cut the following summer.

I felt it was safe to remove the large root as there were several smaller roots growing from the same point at the bottom of the trunk. If these roots weren't visible, I would have waited until I repotted the tree to remove any roots.

Removing a large root

After removing the large root

Indicating roots to be cut during summer

Summer Work

The tree responded well to the initial styling work, so much so that it required thinning in summer. In addition to thinning the branch pads, I reduced about half of the foliage on the upper branches to continue the process of incrementally removing them.

Notice that the tree was propped up so it could grow at the intended planting angle. I created a wood frame for the pot and a fence to prevent the soil from falling out. Propping the tree helped the foliage receive adequate light during the growing season so it could fill in evenly.

Before thinning

After thinning

That was it for the summer work. I continued fertilizing through the remainder of the growing season to maintain the tree's health.

Fence holding the soil in place

WINTER WORK—YEAR TWO

Before pruning

After pruning

After creating additional deadwood features

One year after the initial cutback and pruning, it was time to remove the remaining foliage near the top of the tree, finish the deadwood features, and refine the silhouette by wiring the fine branches.

After completing the work on the trunk and branches, my last step was to repot the tree. I followed the basic repotting sequence and found field soil in the interior of the root ball from when the tree was growing in the ground. After I removed as much of the field soil as possible (about half of the total), it was time to remove the long root I had shortened the previous year.

Long root

Shortening the long root

Repotting complete—front

Left side

Right side

Back

Although the tree now looks more like a bonsai than it did when work started the previous year, the work is far from done. In the coming years, I plan to make incremental improvements by refining the deadwood and the branch structure to give the tree a greater appearance of age. Going forward, I expect the development to be less dramatic but more rewarding as the tree begins to take on the characteristics that signify time in training: a mature silhouette, dense branch pads, and aged deadwood.

Shimpaku juniper—eighteen months after initial work

TOMMY

ACKNOWLEDGMENTS

Thanks to David Fenton for creating the beautiful images featured in the book and to Adam Toth for helping on photo days and preparing the bulk of the trees for photography. Thanks to Stephen Silberblatt, Yuri Aono-Tiede, Christian Werk, Cesar Ordoñez, and Max Vally for additional help preparing trees and helping keep the garden in shape. Thanks to Andrej Kiska for the conversations about setting up a bonsai garden that led to the premise of this book.

At Ten Speed Press, thanks to Lisa Regul for organizing the team that produced the book and to editors Kim Keller and Zoey Brandt for seeing the project through to completion. Thanks to Isabelle Gioffredi for the truly beautiful design, to Jane Chinn for the production work, and to everyone on the production team and in PR, Marketing, and Sales for their contributions.

Thanks to the many readers who provided feedback, including Nick Aleshin, Ian Baker, Lani Black, Sally Cotter, Lukas Fletcher, Sue Heise, Yannick Huot, Austin Jones, Nick Richards, David Ruth, Edward Smyth, David Starman, and Cheryl Sykora. Special thanks to Dylan Ferreira, Nils Schirrmacher, and Eric Schrader for reviewing the complete manuscript. Thanks to Rob Martelli for taking a break from bonsai work to model for a photo (see page 130). Thanks to my dad for facilitating the photo shoots and to my mom for her ongoing support. Special thanks to Haruka Takahashi for her cheerful feedback (and snacks!) and to Lauren Takahashi for help with the planning, writing, and editing of the manuscript, and, more importantly, for being there for me from start to finish.

Thanks to the bonsai professionals who provided their insights for the vignettes and to Christopher Bently (see page 138), Eric Schrader (see pages 40, 42, 76, and 168), Jeff Stern (see pages iv, 18, 19, 34, 39, 41, 43, 147), and Peter Tea (see page 35) for contributing trees for photography. Finally, heartfelt thanks to Daisaku Nomoto for the fantastic work on the juniper featured in Chapter 10.

Opposite: Cesar Ordoñez

INDEX

D

E

F

G

H

I

J

K

L

M

ABOUT THE AUTHOR

Jonas Dupuich runs a bonsai nursery in Northern California, where he teaches and writes about bonsai. He has been growing bonsai for more than thirty years and is the author of *The Little Book of Bonsai* and the *Bonsai Tonight* blog, a weekly publication featuring how-to articles and photographs of bonsai around the world.

In 2022, Jonas cofounded the Pacific Bonsai Expo, a juried exhibit that features beautiful bonsai from around the country. His trees have been selected for display in local and regional exhibits, including the Pacific Bonsai Expo and the US National Bonsai Exhibition. Jonas grows a variety of different species and specializes in developing black pine bonsai from seed. Learn more at bonsaitonight.com.

TEN SPEED PRESS
An imprint of the Crown Publishing Group
A division of Penguin Random House LLC
tenspeed.com

Typefaces: TIGHTYPE's Moderat and Monotype Studio's Albertus Nova by Berthold Wolpe, digitalized by Toshi Omagari

Additional photo credits:
Andrew Robson, page 29
Jonas Dupuich, pages 32, 33, 130, 140, 161
Michael Hagedorn, page 127
Boon Manakitivipart, page 195

Tree featured opposite title page: European beech

Library of Congress Cataloging-in-Publication Data is on file with the publisher.

Hardcover ISBN 978-1-9848-6277-8
eBook ISBN 978-1-9848-6278-5

Printed in China

Acquiring editor: Lisa Regul
Project editor: Kim Keller and Zoey Brandt
Production editor: Serena Wang
Designer: Isabelle Gioffredi
Art director: Emma Campion
Production designers: Mari Gill and Mara Gendell
Production manager: Jane Chinn
Prepress color managers: Zoe Tokushige, Hannah Hunt, and Nick Patton
Copyeditor: Lisa Theobald
Proofreaders: Hope Clarke, Barbara J. Greenberg, and Andrea Peabbles
Indexer: Jay Kreider
Publicist: Jana Branson
Marketer: Allison Renzulli

10 9 8 7 6 5 4 3 2 1

First Edition